D0239680

PRESOWING IRRADIATION OF PLANT SEEDS

Ag TT 85-8-0384

Presowing Irradiation of Plant Seeds

[*Predposevnoe Obluchenie Semian Sel'skokhoziaistvennykh Rastenii*]

Second Revised and Enlarged Edition

N.M. Berezina and D.A. Kaushanskii

Editor
A.M. Kuzin

Translated from Russian

Amerind Publishing Co. Pvt. Ltd., New Delhi
1989

Atomizdat Publishers
Moscow, 1975

Translated and published under an agreement for the
United States Department of Agriculture, Washington, D.C.,
by Amerind Publishing Co. Pvt. Ltd., 66 Janpath, New Delhi 110 001

Translator: Dr. A.K. Dhote
General Editor: Dr. V.S. Kothekar

Printed at Gidson Printing Works, New Delhi, India

UDC 63:621.039.8

The book reviews published data and presents results of studies conducted by the authors on the presowing gamma-ray irradiation of seeds of agricultural crops. The authors describe the theoretical basis of the stimulatory effect of small irradiation doses and the mechanism of their action. It has been shown that it is possible to raise the yield of agricultural crops, accelerate ripening and improve the quality of raw material obtained from irradiated seeds. The book includes results of field experiments and production trials as well as results obtained in the first year (1973) of introduction of presowing irradiation.

The second edition (first published in 1964) has been supplemented with information on radiation technology used in irradiation of seeds under production conditions.

The book is meant for radiobiologists, specialists in agriculture, workers in radiation safety as well as for students and post-graduates of agricultural institutes and biology faculties.

The book is richly illustrated with 53 figures, includes 137 tables and has a bibliography of 364 citations.

Preface

Since the publication of the first edition of this book [21] vast changes have taken place in the research on presowing γ-irradiation of agricultural crop seeds. Vast experimental material has accumulated, which has made it possible to identify in radiobiology an independent branch, agricultural radiobiology.

Contemporary knowledge about the mechanism of the effect of small doses of irradiation at the molecular, cellular and organism levels testifies about the complex series of sequential physiological and biochemical changes which arise during seed irradiation and become visible from the moment of seed germination to the maturity of the new crop.

At present, presowing irradiation of seeds can be considered a novel agronomic measure which enables us to harness atomic energy in the form of ionizing radiations to increase the yield potential. Numerous results testify about the biochemical changes that improve the quality of raw material by virtue of its large content of such valuable metabolites as proteins, fats, carbohydrates, vitamins and so on. It has also been shown that upon irradiation new quantitative changes do not occur in the biochemical composition of seeds or in the plants raised from them, because all changes are a result of an increase in the content in the precursors of those metabolites whose synthesis has been genetically predetermined for a particular plant species. This fact, confirmed by exhaustive, reliable research over many years by many scientists, guarantees the safety of the stimulating doses of irradiation. The possibility of improving the quality of the precursors through gamma-irradiation increases the significance of this measure.

Due to the gamma-irradiation the maturity is enhanced of crops such as strawberry, watermelon, muskmelon, radish, cucumber and cabbage. For growers this is more important than yield increase, because the price of early-maturing produce is much higher at the market. Presowing irradiation of seeds enables us to obtain additional produce without changing the agronomic practices developed for reaping high yields and without increasing expenditures on labor.

In the Special Design Bureau (SDB) of the Institute of Organic Chemistry of the USSR Academy of Sciences, D.A. Kaushanskii developed a self-propelled field unit 'Kolos' to irradiate seeds. It has a capacity for one ton seeds per hour at a total dose of 1 c. This helped launch large-scale production trials on presowing irradiation. During 1968–1971 the Kishinev Agricultural Institute, named after M.V. Frunze, conducted production trials on the Kolos unit and the treatment of presowing irradiation for an area of about 12,000 ha after an agreement with the USSR State Committee for the Use of Atomic Energy. It was done with the consultation and methodological assistance of the Institute of Biophysics of the USSR Academy of Sciences and the SDB of the Institute of Organic Chemistry of the USSR Academy of Sciences. Similar trials were begun in 1971 in the Pavlodar oblast of the Kazakh SSR where 5,000 ha were sown with irradiated seeds. In 1972, the efficacy of presowing irradiation of seeds was tested on an area of about 70,000 ha in Moldavia, Kazakhstan and Kirgiz.

From the results of these production trials, it was decided to widely extend the presowing irradiation of maize seeds in the collective and state farms in the above republics from 1973. The recommendations on the introduction of presowing irradiation and further expansion of work in this direction are now quite opportune, because increasing the yield potential of agricultural crops is one of the top priority tasks assigned to the planners of our country's economy by the Party and the government.

The production trials of presowing γ-irradiation of seeds conducted in the USSR attracted the attention of radiobiologists abroad. In 1968, a consultative council was established in Vienna to solve the problem of the practical use of γ-irradiation of seeds. A comprehensive report on the state of the art of the research on presowing irradiation of seeds in the USSR and abroad was made by Glubrecht (Hanover) [288]. While reviewing the research on presowing irradiation of seeds from a historical perspective, Glubrecht asserted that Soviet radiobiologists are initiators of research in this direction. In the post-war period, ionizing radiation was used abroad mainly to obtain radio-mutations [261] and no efforts were made to use small doses of radiation to stimulate because of the poor reproducibility of favorable effects. In the USSR, however, during this time, in-depth studies were made to understand the mechanism of the action of small doses of radiation and to reveal the reasons for the transient nature of their effects. According to Glubrecht's data, researches on the use of stimulating doses of radiation have received great impetus during the past few years. Favorable results have been reported for certain crops in Canada, Poland, Hungary, Czechoslovakia, the German Democratic Republic, the Federal Republic of Germany, France, Bulgaria and other countries. Institutes of agricultural radiobiology have been established in a number of countries: in Wageningen (The Netherlands), in Jülich and Munich (FRG) and so on. A special international journal, *Stimulation Newsletter*, devoted to the study of the effects of radiostimulation has been published since 1971 [28].

Simultaneously, in the USSR, experiments were conducted on the development of new technology to irradiate seeds. The work of radiobiologists, physicists, design engineers and extension agronomists was directed toward the solution of a common problem, viz., the use of atomic energy to increase the yield potentials of our fields. This task has been accomplished with production trials on presowing irradiation and on the Kolos unit.

The authors of this book aimed to make a historical review of the research on presowing irradiation of seeds using γ-radiation and identify possible ways to use this as an agronomic measure.

In view of the introduction of presowing irradiation of seeds in agriculture, a chapter on radiation technology has been included in the second edition of this book for the edification of agronomists and biologists.

The preface to the second edition and Chapters 1–4 have been written by Doctor of Biological Sciences, N.M. Berezina, and Chapter 5 by Candidate of Technical Sciences, D.A. Kaushanskii.

Contents

1. Historical Review of Researches on the Effect of Ionizing Radiation on Seeds

Röntgen's discovery of unknown rays in 1895 drew the attention of scientists of the most diverse branches of knowledge, including biologists who made efforts to discover the effect of these rays on living organisms. The first researcher to study the effect of x-rays on vegetative objects was Schober [346]. He subjected air-dried oat seeds to small doses of x-rays and established the increase in the germination percentage and energy of germination of the irradiated seeds. Almost analogous results testifying to these stimulating effects of x-rays were obtained by Maldinay and Touvenin [321] and Schmidt [345], who observed the increase in germination and energy of germination of the irradiated seeds of lettuce, millet and lesser bindweed.

Numerous experiments to study the effect of radium radiation on seeds of different agricultural plants were conducted by Geiger [281–287]. In individual experiments he observed clearly manifested stimulation effects in the increase of germination, energy of germination and development of seedlings raised from the irradiated seeds. On experiments with wheat seeds he noticed that at first the emergence and development of seedlings were delayed as a result of the irradiation, but later these seedlings quickly surpassed the control seedlings in their growth. Geiger did not note this stimulation in irradiated barley or lupin seeds. Observing significant deviations in the behavior of irradiated seeds from experiment to experiment, the scientist came to the conclusion that ionizing radiation exerts complex effects covering various aspects of plant development. He was the first scientist to begin a study on biochemical and physiological changes in irradiated seeds and their seedlings. He also was the first to make a comparative evaluation of α-, β-, γ- and x-radiations. As a result, Geiger concluded that x-radiation is the most effective regarding its biological action.

Simultaneously with Geiger, the German scientist Köernike also extensively studied the effects of x-radiation on seeds [302–306] of beans, corn, lupin and others. While studying the effect of x-radiation, he took into account the germination capacity and germination vigor of the irradiated seeds and conducted cytological studies. For the first time, he noticed the dissimilar radiosensitivity of different species of seeds (according to his data, radish was found to be the most radiosensitive). Observing the development of seedlings raised from the irradiated seeds, Köernike noticed that the stimulation, which appeared in the increase of germination and energy of germination, disappeared during further development. Köernike's concept on temporary stimulation effects was presented in his theory of intermittent stimulation.

However, the research of Soviet scientists has shown the weakness of this theory. Particularly, N.M. Berezina established that when cucumber seeds were irradiated with a dose of 0.1 c, plant growth was stimulated initially, but the further growth of these plants was no different from that of control plants, which confirmed Köernike's theory. However, with much higher doses of irradiation (0.3 c), the stimulation continued to the end of the vegetative growth and led to an increase in the green yield (by 23%). This example shows that the disappearance of the stimulation effect in ontogenesis was the result of the wrong selection of stimulating dose in Köernike's experiment.

Evler [279] confirmed the stimulating effect of irradiation on the germination and energy of germination of radish, lettuce and pumpkin seeds. He noticed the particular effectiveness of irradiation on radish seeds.

Promcy and Drevon [338], working with lentil, rye, white lupin and bean seeds, showed that the effectiveness of small dose of radiation increases if the growth of the irradiated plants takes place at an increased temperature. The x-radiation dose used by the authors at increased temperature always exerted a favorable effect on the germination of seeds and their subsequent development.

With the perfection of experiments on seed irradiation the researchers switched over from observations on germination and energy of germination to observations on plant growth and development during the entire vegetative period. Phenological observations of plants raised from irradiated seeds showed their much earlier development. Guilleminot [290–293], while growing irradiated seeds under field conditions, noted an increase in yield along with enhancement of the plants' maturity. The experiments of Schwarz [348], Miege and Coupe [327] confirmed that presowing irradiation of seeds in a definite dose stimulates the growth and development of plants and increases their yield potential. Schwarz suggested the possibility of using presowing irradiation of seeds to increase yield potential. Köernike repeated his experiments but did not get similar results and thus doubted the stimulating effect of small doses of irradiation and its practical application.

In 1915, Köernike published a review of the research on the effect of small doses of irradiation that aroused sharp discussions and controversies which are cited even today. The basis of the discussion was the inability to always reproduce the exact data of Köernike and other radiobiologists [306].

Many research projects appeared to negate the possibility of stimulation through the effect of irradiation.

For example, while testing the effects of radium radiation for 24 hours on dry seeds, Bequerel [269] did not observe any deviations from the control but, when they were irradiated for one, then two weeks, he noticed an increasing depression in the growth of the seedlings. As a result of these experiments he concluded that there were only damaging effects from seed irradiation. However, it is not true, because it is quite possible that an intermediate dose from the first days to two weeks might have been the stimulating dose. S.V. Gol'dberg, who was the first Russian researcher in this field to use high doses of irradiation in his experiments, noticed that irradiation can only cause damaging effects [77]. Sax [344] while studying the effect of x-radiation on dry seeds, mentioned only the harmful effects of radiation, stating that in all radiation doses studied by him he observed seedlings which were drastically shortened and turned dark brown. At present it is well known that such changes occur only with fatal doses of radiation and never with a stimulating dose.

Hagberg and Nybom [296], irradiating potato tubers with doses of 1, 2, 5–20 c, concluded that the stimulation effect was absent for potato. It is quite natural, because several years of research of V.S. Serebrenikov [213, 214], V.T. Parfenov [180], V.T. Parfenov and N.M. Berezina [181], N.M. Berezina and R.R. Riza-Zade [38] showed that the stimulating dose for potato is 0.1–0.5 c (depending on the variety).

In experiments which negated stimulation, it was not always possible to reveal the reason for the absence of stimulation or its non-reproducibility which had agitated the radiobiologists. They tried to explain these phenomena. For example, Uber and Goodspeed [361] believed the reason for poor reproducibility was that different researchers based the stimulation effect on different indices. The stimulation of germination and energy of germination could have been caused by one dose of radiation and the stimulation of growth associated with the rate of cell division by another dose.

From our point of view, among all researchers Geiger was closest to understanding the real position. He considered that the stimulatory effect was caused not by the dose, but by complex series of biological phenomena accompanying radiation. Iven [297] irradiated dry and soaked bean seeds and established the stimulation effect and conducted much deeper research into the change in effect of the same stimulating dose depending on the moisture content of the irradiated seeds.

Patten and Wigoder [335], Altman, Rochlin and Gleichgewicht [262] through several experiments on agricultural crops showed the stimulation of growth, development and enhancement of maturity of plants grown from irradiated seeds. L.V. Doroshenko [93] sowed oat, barley, rye, millet and flax seeds in the field after their x-irradiation and established the enhanced germination of rye and oat seeds as well as other phases of their development, viz., tillering, flowering and ripening. He was not able to establish this for millet and flax.

I.B. Chekhov [246], who worked in the Tomsk University, irradiated rye, oat and barley seeds and noticed the activating role of small doses of x-ray radiation on the growth and development of plants. G. Frolov [237], while irradiating wheat seeds with 0.125–1 c doses, found a sharp increase in the energy of germination, considerable reduction in the vegetative period and up to 60% increase in the grain yield.

As a result of irradiation, B.T. Tsuryopa [242] obtained enhanced ripening and a 31% increase in the yield of oats when compared with the yield of seeds without irradiation. In this researcher's experiments with flax seeds the stimulation was observed only at early stages of development, but later the plants raised from irradiated seeds lagged behind the control plants in their growth. B.I. Tsuryopa noticed only radiation's depressing effect on irradiated soybean seeds.

Long and Karsten [317] conducted experiments on irradiating soybean and got the opposite results. While irradiating the seeds of Manchurian and black-seeded Wilson soybean, they observed a distinct stimulation expressed by very vigorous tillering and branching in plants grown from the irradiated seeds.

MacKey [319, 320] obtained a considerable increase in the yield of barley and beans as a result of small dose x-ray irradiation of seeds. He patented the irradiation procedure for bean seeds as an agronomic measure to facilitate an increase in their yield potential.

Irradiating the seeds of wheat and some other crops Jonson [299] noticed an accelerated growth and development of wheat.

In the USSR during the thirties, a number of studies were published, which were conducted at the initiative of L.P. Breslavets [51–53, 57, 59], A.S. Afanas'eva [15–18] and A.I. Atabekova [11–14]. These researches confirmed the optimal stimulating effect of x-radiation in the dose of 0.25 c on rye and in the dose of 0.35 c on peas. The importance of these studies lies in the fact that the experiments were conducted under conditions of correct dosimetry or radiation monitoring. The research work of this period was summed up by L.I. Breslavets by the release of the book *Rasteniya i Luchi Rentgena* (Plants and x-rays) [54] which played a significant role in the development of a study on the effect of radiation on plants in the USSR. From this time scientists began much deeper studies on the effect of small doses of radiation. During our further researches it has been established that the stimulation effect changes, depending not only on the dose, but also on the physiological condition of the seeds and a whole series of environmental factors like humidity, temperature, etc.

Simultaneous with the development of research on seed irradiation, scientists began to conduct experiments with seeds of agricultural crops soaked in the solutions of radioactive substances and the application of these substances as fertilizers in the soil. In these methods the radioactive substances are absorbed by the seeds; remain inside the seeds and later in the plants. These substances act as sources of the same ionizing irradiations present in the object being studied.

Research in this direction was initiated in the Paris Academy of Sciences. In 1912–1914 a special journal, *La Radium Culture*, was published in France. Since 1915 studies on the effects of radiation on plants began to expand and in 1925 a special Institute of Radioactive Fertilizers was set up.

In the USSR experiments on soaking seeds in radioisotope solutions and the application of radioactive fertilizers were conducted in the Ural branch of the USSR Academy of Sciences by N.V. Timofeev-Resovskii [225–228], N.A. Poryadkova [186–189], N.V. Luchnik [149, 150]. In this very direction research was conducted in the Institute of Plant Physiology of the UkrSSR Academy of Sciences under the guidance of P.A. Vlasyuk and associates [65–70, 72, 73]. Among the studies on the application of radioactive fertilizers conducted in other research institutions, it is worthwhile to mention the research of M.G. Abutalybov and N.B. Vezirova [2], D.A. Agakishev [4], P.R. Borodin and associates [49], N.B. Vezirova [62], A.A. Drobkov [94, 95] and N.G. Zhezhel' [97–101].

Without dealing with the individual experiments conducted by the research workers of the Ural branch of the USSR Academy of Sciences, we shall mention that by 1957, of the 42 experiments with seeds of many different species of agricultural plants soaked in radioisotope solutions of weak concentrations (0.25–1 mc/1) most of the crops sown over an area of about 300 ha gave 5–20% more yield compared with the control crops.

Numerous experimental data testifying to the effectiveness of soaking seeds in solutions with the radioactive fertilizer mixtures obtained at the Institute of Plant Physiology of the UkrSSR Academy of Sciences were confirmed in tests on the effect of radioactive fertilizers under field conditions.

However, neither the method of soaking seeds before sowing nor the method of applying radioactive fertilizers found wider acceptance in practical, large-scale farming. The reason was that working with radioactive elements during seed soaking or their application in the soil can be dangerous not only to the persons working with them, but for the entire population. Apart from this, in all cases of applying radioactive substances (soaking seeds in their solutions or their application in the soil) there is an inevitable risk of retaining residual radioactivity in the yield. This has often been noticed by V.M. Klechkovskii [136] and others.

But presowing x- and gamma-ray irradiation has no radioactivity danger—neither for those working with the irradiated seeds nor for those consuming plants grown from them. By virtue of this, presowing irradiation is being thoroughly and extensively studied in the USSR in experiments as well as in the fields.

Lately, research on the presowing irradiation of seeds has been stopped in the West European countries. France is an exception where the Commissariate on Atomic Energy supported and financed these works conducted under the guidance of P. Vidal.

In the USA, England and Sweden radiobiological research in a different direction was taken up extensively [322]. Their main aim was to obtain radiomutants of different agricultural plants.

During the fifties long-term experiments on the effect of chronic irradiation were initiated in the USSR and other countries. So-called gamma fields were extensively used as a method to study chronic irradiation. A gamma-field is a plot of experimental land with a source of radiation installed in the center. In working condition it ascends above the field level. During agricultural experiments in the gamma-field, the source is lowered underground to ensure complete safety to the workers. The first gamma-field was established in Brookhaven in 1952. The source of radiation was ^{60}Co with an activity of 1,800 c.

The first gamma-fields in the USSR were established at the Institute of Biophysics of the USSR Academy of Sciences (1954–1957) and at Gorkii University (1956). In these gamma-fields the source of radiation was ^{60}Co with an activity of 1 c. Small source activity made it possible to study chronic effects of only a small stimulating dose of ionizing radiations.

In 1956 a gamma-field was set up under the All-Union Research Institute of Fertilizers and Agropedology at Barybino (Moscow province). The source on this field was ^{60}Co with activity of 225 c. The plants were planted in the gamma-field at equal distances from the source of radiation. The cumulative dose of radiation changed proportionately to the square of the distance from the experimental plants to the radiation source. The entire area of the experimental field of the Institute of Biophysics of the USSR Academy of Sciences was divided into four zones situated radially around the source; the strength of the dose in the first zone under the source was 1.8 r/hour, in the second zone—0.552 r/hour, in the third—0.108 r/hour, and in the fourth—0.023 r/hour, which made it possible to obtain a cumulative dose of radiation in the different zones from 340 to 4.3 r.

Such a construction enables us to study the effects of different doses of radiation for different durations of time as well as to compare the effect of chronic radiation received by the plant during the entire vegetative period with the effect of equal doses of radiation received in the form of single, short or, as it is called, acute irradiation (Fig. 1.1).

The effects of stimulating doses of chronic irradiation on corn, buckwheat, carrot, strawberry and sugarbeet were studied in the gamma-fields of the Institute of Biophysics; the results have been presented in the works of

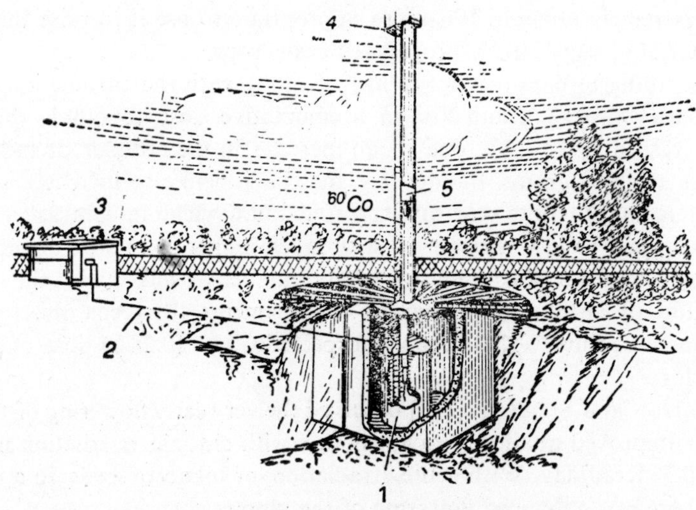

Fig. 1.1. Schematic diagram of the unit for the irradiation of plants in the gamma-field:

1—container for the storage of the radiation source;
2, 4—systems to raise and lower the source; *3*—unit control;
5—protective cover of the container.

L.P. Breslavets, N.M. Berezina, G.I. Shchibrya, M.L. Romanchikova [39, 55, 58]. The experiments thus conducted showed that with corresponding doses of chronic irradiation as well as acute irradiation the stimulation effect emerges and is expressed in increased yield potential and in the improved quality of plants grown on the gamma-field. For example, buckwheat came to flowering three to four days earlier, gave up to 42% greater yield of green matter when it was grown to obtain rutin; the rutin content in it increased up to 18% in comparison with the control plants. Corn grown on the gamma-field flowered and matured earlier and many additional cobs emerged as a result of irradiation. In this case the yield of green matter increased by 18–23% in comparison with the control (Fig. 1.2).

When strawberries were grown on the gamma-field, the berries ripened three to four days earlier than in the control and the yield of the berries increased up to 40%. Apart from commercial plantations of strawberry, mother plantations were also irradiated in the period of runner formation. After planting runners in the commercial plantations the berry yield was recorded for two years. The results showed that when runners were irradiated with doses of 0.34–0.14 c in the vegetative period, the berries ripened faster, while the yield corresponding to the dose increased by 29 and 33% in comparison with the control. The reproducibility of the stimulation effect was considerably better with chronic irradiation than with acute irradiation. However, in spite of this, extensive practical application of chronic irradiation is extremely difficult. It is more appropriate to use it to raise the planting material of vegetatively propagated cash crops.

According to data of Glubrecht [288, 289], with the chronic irradiation of the barley variety Braun Visa in a cumulative dose to 500 r, the grain yield increased by 13–23% due to an increase in the number of spikes and the mass of 1,000 grains. Besides this, the chaff yield also increased sharply. It is interesting to note that an increased chaff yield in contrast with the increased grain yield was observed with much higher (3 c) dose of radiation. Kervegni also got a sharp increase in the number of spikes with the chronic irradiation of wheat. In Gonkarnev's experiments the green mass yield of clover increased by 70% with an unexpectedly low (0.24 r) dose of chronic irradiation.

Sparrow and Singleton [353] observed the very early flowering of tobacco and the improved quality of raw produce with chronic irradiation in doses of 0.1–0.3 Krad/day. With acute irradiation of tobacco seeds in a dose of 1.8 c, there was enhanced flowering of the plants.

a b

Fig. 1.2.—Multicob corn grown on gamma-field:
a—control; *b*—irradiated plant.

In Hungary, Balint (1968), Menyhert and associates [336] irradiated seeds of two varieties of peas, viz., Pti provancal' and Kal'vedon. The plants were grown in phytotron; the optimal stimulation effects from radiation appeared with doses of 0.75 and 1.5 c. In 1968, experiments were conducted there on the treatment of potato tubers with ^{60}Co gamma-rays in doses of 0.1–0.3 Krad. In this case the potato yield increased by 44% and this increase in yield was due to the increased number of tubers in a plant. The

Table 1.1. Grain yield of spring barley from the plot for presowing irradiation of seeds, % (experimental station: Laken)

Barley variety	Year	Control, g	Radiation dose, rad			
			1	10	50	100
Early	1964	—	—	—	—	—
	1965	32.3	98	100	96	105
	1966	22.5	113	111	116	122
	1967	33.2	110	104	107	99
Union	1964	37.6	102	104	101	100
	1965	26.8	104	109	107	102
	1966	24.8	112	103	106	102
	1967	29.1	113	121	121	133

Table 1.2. Results of presowing irradiation of seeds of some agricultural crops in Canada

Crop	Year	Dose, rad	Results
Sweet corn	1968	100–1,000	Increased yield by 34–41%
Lettuce	1968	100–1,000	Increased yield by 35–50%, with much earlier ripening.
Tomato	1969–1970	300	Increased total yield by 21–27%
Peas	1968	100	Increased number of flowers by 15%. Yield of early pods increased by 90%, total yield by 24%
Pumpkin	1968	100–1,000	Early harvest exceeded the control by 37–41%, total yield by 15–30%
Cucumber	1969	300	Early harvest exceeded the control by 40–90%
Wheat	1970	1,000	Increase of yield by 15%
Barley	1969	1,000	Slight increase of yield
	1970	100–1,000	-do-

ascorbic acid content also increased in the tubers as a result of irradiation. Besides these, experiments were conducted in Hungary on the presowing irradiation of corn, winter and spring barley, sugarbeet and tomato seeds (in the spring and fall harvests). Analogous experiments with peas were conducted by the Japanese scientist Komuro [312].

In Bulgaria, Panasov irradiated water-melon seeds in the dose of 10 c, and the fruit yield increased by 15% and the sugar content by 8% in comparison with the control. Experiments conducted in the FRG by Sus [356–358] are of great interest. This radiobiologist used an extremely low dose for the presowing irradiation of seeds. Generalizing his results, Sus showed that, based on averages from four years of data, low doses of radiation like 1–10 rad also give significant increases in yield, just like much higher doses (Table 1.1). Here Sus observed not only quantitative, but also qualitative changes in the raw produce grown from irradiated seeds [358].

At the Fourth International Conference on the Peaceful Use of Atomic Energy, held at Geneva in 1971, a number of reports were presented about the stimulation effect of low doses of radiation. MacKey presented the results of experiments conducted in Canada for three years, according to which, for doses of 0.1–1 c at the strength of 40 rad/min for most of the crops studied, he obtained stimulation with regard to the following indices—germination, energy of germination, flowering, maturity and yield (Table 1.2).

2. Radiosensitivity of Plants and its Change

Maximum radiosensitivity is possessed by warm-blooded animals, for whom the lethal dose varies within a range of 0.4–1.2 c. Minimum radiosensitivity is possessed by microbes and bacteria which perish at doses of 600–2,000 c. An intermediate position regarding radiosensitivity is occupied by vegetation for which the lethal dose varies within a range of 10–500 c. The comparative radiosensitivity of different living organisms is given below:

	Lethal dose, c		Lethal dose, c
Guinea pigs	0.4	Dry seeds	10–500
Mice and rats	0.8	Molds	200–500
Rabbits	1.2	Microbes and bacteria	600–2,000
Frogs	1.8–2.0		

It is essential to note that the amplitude of variations of lethal doses in plants is considerably higher than in animals and microorganisms.

Apart from differences in radiosensitivity observed in living organisms at different stages of their evolutionary development, we find variations of individual radiosensitivity, i.e., the change in radiosensitivity of individual tissues of the same organism and its change under different physiological conditions of the same organism and so on. The reasons for the different radiosensitivities of living organisms, including plants, have still not been fully explained, despite many theories and hypotheses attempting to explain this phenomenon, which is one of the central problems of modern radiobiology.

The change in radiosensitivity is of very great significance for agricultural radiobiology because this change, linked with the effects of environmental factors and the physiological condition of the plants, is one of the basic reasons for the poor reproducibility of the stimulation effect. Therefore, the study of the mechanism of the radiation effect and reasons for the changing radiosensitivity of plants is a top priority problem of agricultural radiobiology.

E.I. Preobrazhenskaya [191] compiled a comprehensive review of studies on seed radiosensitivity. She generalized data about the radiosensitivity of a very large number of dormant seeds representative of two sections, two classes, 42 orders, 63 families, 262 genera, 508 species and 208 botanic varieties and sub-varieties. The entire range of plants studied by her was divided into three groups: plants which perish at 10–25 c doses are categorized as highly sensitive; plants which perish at doses of 25–100 c are categorized as moderately radiosensitive; and the plants that perish at doses of 100–300 c and above are categorized as least radiosensitive. Among E.I. Preobrazhenskaya's conclusions of great interest for us is the establishment of a correlation between the radiosensitivity of plants and the stimulation dose. The most stable effect of stimulation was observed by her in the group of radioresistant plants and the least stable effect of stimulation in the radiosensitive plant group.

T.E. Guseva [83] established that the range of stimulation doses is narrow for radiosensitive plants, but significantly wider for radioresistant plants. Using E.I. Preobrazhenskaya's table, we can establish the range of stimulation doses of radiation for crops being studied afresh.

Different plant radiosensitivities were mentioned by several radiobiologists. Köernike in 1904 [302, 305], while irradiating seeds of different agricultural crops with a wide range of doses, found the lowest radiosensitivity among cruciferous plant seeds, particularly radish. According to his data, the intermediate position was occupied by representatives of the family Graminae.

Shwartz, Czepa, Kol'tsova, Nadson and Zholkevich in their research pointed out different radiosensitivities of plants within the same species. Shwartz, et al., saw this as one reason for the non-reproducibility of the stimulation effect. The efforts to equalize radiosensitivity differences by selecting seeds based on size did not produce the desired results. Later Jonson [301] conducted experiments to study the radiosensitivities of different species of agricultural plants. Under comparable conditions, she irradiated seeds of more than 100 species of plants with a wide range of doses. Analogous experiments on a somewhat smaller scale were conducted by L.P. Breslavets and associates [51], Gustafsson [294, 295] and Nybom [332]. Gustafsson considered that seed radiosensitivity depends on their oil content; however, we did not find such a correlation in a number of oilseed crops. According to T.E. Guseva's data [83] in mustard seeds, the least radiosensitive plant at 40–42% oil content, the LD_{100} = 750 c, but in soybean seeds whose LD_{100} = 40 c (19 times less) the oil content is almost the same—35–45%.

However, N.M. Berezina and T.E. Guseva [33, 34] established a direct correlation between the radiosensitivity of oil crop seeds and their unsaturated fatty acid content (Fig. 2.1).

Sparrow and associates [352, 353], working with a large number of plant species, tried to find a correlation between chromosome number (or volume of the nucleus) and radiosensitivity. However, they were only able to establish a direct correlation within the limits of individual, artificially selected

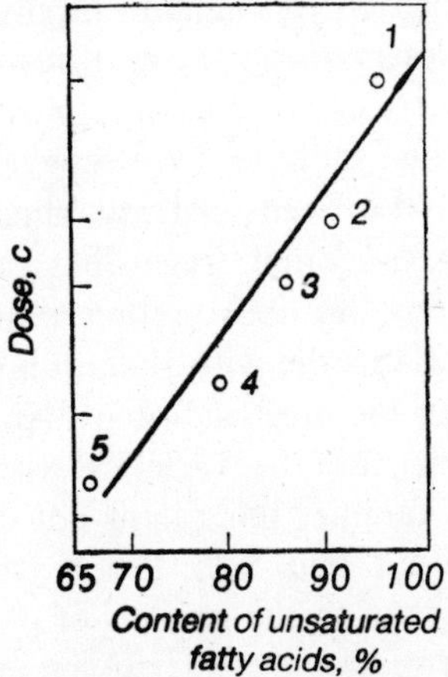

Fig. 2.1. Correlation between the content of unsaturated fatty acids in the seeds of some oilseed crops and their radiosensitivity (values of doses are given along the ordinate axis—40, 80, 200, 300, 750 c):
1—mustard; *2*—castor; *3*—sunflower; *4*—peanut; *5*—soybean.

groups of plants, termed by them as radiotoxic. Saric [343] did not find a direct correlation between the radiosensitivities of different plant species and the chromosome number:

Plant species	Chromosome number	Radiation dose, c
Trifolium repens	16	30.0
Pisum sativum	14	8.0
Phaseolus vulgaris	22	10.0–12.5
Lupinus luteus	24	15.0
Glicena soya	22	15.0
Linum usitatissimum	40	7.5
Papawer somniferum	32	20.0–60.0
Canabis sativum	20	7.5
Brassica napus	38	90.0
Sinapis alba	18	92.0
Helianthus annus	34	5.0
Poa pratensis	72	10.0–15.0

In spite of a great deal of research to reveal the reasons for different plant radiosensitivities, this problem still remains unsolved. Diverse explanations for this phenomenon are given by different scientists.

N.V. Timofeev-Resovskii and N.A. Poryadkova explained the different plant radiosensitivities by the size of the seeds. They suggested that bigger seeds with large surfaces must be more radiosensitive than smaller seeds. However, the presumption put forth by them cannot be considered a general rule. This rule is confirmed in the legume family. The horse-bean, whose seeds are the largest, belongs to the group of most radiosensitive plants for whom the lethal dose is 10 c. Sweetclover, clover and lucerne, with the smallest seeds, are the least radiosensitive representatives of this family for whom the lethal dose is 300 c. Seeds of lentil, chick pea, green gram, lupin and vetch are of average size and, regarding radiosensitivity, occupy an intermediate position among the above representatives of the Leguminaceae family. Seeds from these plants die with radiation doses of 25–50 c. Among different varieties of peas the most radiosensitive are the large seeded ones and the least radiosensitive are the varieties with smaller seeds. However, within the limits of other families this rule is not confirmed.

N.V. Timofeev-Resovskii and E.I. Preobrazhenskaya put forward a hypothesis to the effect that, in the process of plant evolution, their radiosensitivity changes and evolutionally older plant species are more radioresistant than the evolutionally young ones.

Barton's hypothesis [265–267] that the different radiosensitivity of individual species and groups of plants is due to the anatomical structure of the seeds deserves attention. Barton attributes the low radiosensitivity of seeds of the Cucurbitaceae family to the fact that, below the compact fruit membrane, there is a semitransparent, thin, gas-impermeable membrane. Due to the presence of such a membrane, the embryo and cotyledons become anaerobic after irradiation, which considerably reduces the harmful effects of radiation, thereby providing low radiosensitivity to the seeds.

He related the slow germinating seeds with small embryos to the group of highly radiosensitive seeds; and they germinate slowly. Therefore, at high radiation doses the embryos of these seeds develop for a long period of time on nutrient reserves of the endosperm changed by the radiation. From our point of view, Barton's research is of great interest because it reveals different causes for individual plant radiosensitivities.

Sparrow, et al. [352] investigated the radiosensitivity of 230 species of plants which were grown in the gamma-field. They established that radiosensitivity, determined by acute and chronic irradiation, is uniform. However, in order to get the same effect acute irradiation must be 13 times higher than the chronic irradiation which is received by plants over a period of 12 weeks.

While comparing the nuclei volume in somatic and generative cells, the authors noticed that somatic cells have much smaller nuclei and they are less sensitive than the generative cells whose nuclei are considerably bigger.

After determining that there exists a well established, direct correlation between the volume of the nucleus and its DNA content, Sparrow expressed the opinion that the quantity of DNA in the nucleus is the factor controlling different plant radiosensitivities. The hypothesis on the dependence of plant radiosensitivity on the volume of the nucleus became the basis for the presumption that the high ploidy level of plants can be the reason for a decrease in their radiosensitivity. Numerous experiments were conducted in many countries to solve this problem.

While determining the radiosensitivity of diploid and autotetraploid buckwheat, V.V. Sakharov and V.L. Mansurova showed that the autotetraploid is considerably less sensitive. L.P. Breslavets and Z.F. Mileshko noticed a much lower radiosensitivity in the tetraploid rye in comparison with the diploid rye. However, it was not possible to detect a direct and well expressed correlation between the ploidy level of plants and their radiosensitivity in all cases. Saric [342, 343] compared the correlation between the radiosensitivity and the ploidy level of naturally evolved and artificially obtained polyploids. With the irradiation of air-dried seeds from artificially obtained polyploids of the barley variety Opal and the rye variety Petkus, he established the much lower radiosensitivity in the tetraploid in comparison with the diploid forms.

While studying the radiosensitivity of wheat with different ploidy levels ($n = 7, 14, 21$) which evolved naturally, Saric established the similar sensitivity of these species to the effects of x-radiation. However, observations on other species of wheat, viz., *T. monoccocum* and *T. vulgare*, showed that the haploid wheats were more sensitive.

Palenzona [333] studied the comparative radiosensitivity of di-, tetra- and hexaploid wheat and tri- and hexaploid clover. His criteria of the effect of radiosensitivity were plant height, root length and dry mass. The radiosensitivity of objects studied by him reduced with an increased ploidy level.

When comparing hexaploid wheats with tetraploids, Natar'yan established the low radiosensitivity of tetraploids. However, he observed this correlation only during x-radiation. Hexaploids were found to be more resistant to ultraviolet-radiation.

Bora [270] compared the radiosensitivity of two wheat species ($n = 14, 21$). In his investigations the following indices were the radiosensitivity criteria: seedling emergence, plant height and chromosome aberrations. In some indices the radiosensitivity was lower in tetraploids, while in other indices, it was lower in hexaploids.

While subjecting the seeds of tetraploid and hexaploid forms of barley to the effect of neutrons and x-radiation, Konzak and Singleton [315] established that tetraploids were less sensitive to the effect of thermal neutrons, but hexaploids—to the effect of x-ray radiation.

These data testify that radiosensitivity of polyploids is controlled not only by chromosome number, but also by some other indices which are characteristic of different varieties and botanic sub-varieties of polyploids.

Some authors tested the radiosensitivity of hybrids. For example, when irradiating corn seeds, Saric [343] showed that the hybrids are considerably less radiosensitive than the initial parents, but the sensitivity of double hybrids, according to his data, was still lower. One of the parents died at the dose of 12.5 c, another—at the dose of 25 c, but the hybrid obtained from them survived at the dose of 25 c.

In his experiments with corn, Notani [329] confirmed Saric's data. The hybrids were considerably less radiosensitive than the parental forms: in the first series of experiments on irradiation at the dose of 30 c, 90.5% of hybrids and 79.4% of the best parental forms survived. In Notani's second series of experiments on irradiation at the dose of 20 c, the survival rate of hybrids and the parental forms was 97 and 55% respectively. The author considers that, in the hybridization of plants, a great role is played by the biochemical composition of parental plants, particularly the enzymatic activity of each of them. For example, in crossing corn forms with different enzymatic activities, the range of enzymatic activity of hybrids expands with a consequent increase in the resistance of hybrid plants to the changing external conditions, including radiation.

The degree of maturity of the grain being irradiated is also one factor which influences changes in radiosensitivity. While irradiating wheat, barley, oats and rye seeds, Saric observed significant changes of radiosensitivity depending on the grain maturity; immature seeds containing up to 80% water, have underdeveloped embryo and endosperm and were considerably more radiosensitive than mature seeds. One of the reasons for the increased radiosensitivity of immature seeds, according to the author, was their high moisture content. O.S. Engel' [255] irradiated wheat seeds collected in the milky, milk-dough and maturity stages and also showed that mature seeds are least radiosensitive. Cytological analysis showed that there are numerous cell divisions in the embryo during the early stages of seed ripening. The number of dividing cells reduces with ripening. The author considers the difference of mitotic activity in the process of ripening as one reason for the change in radiosensitivity.

Mericle [325] irradiated growing plants during a study of x-radiation effects on the generative organs in ontogenesis. Histological analysis showed that the damaging effect of irradiation was less if it took place before the differentiation of embryo tissues and not after that moment. The plants grown from seeds that were irradiated at the earliest stages of fertilization had undergone numerous morphological changes.

The above data are adequate to presume that the radiosensitivity of individual species or groups of plants changes depending not on any one of the indices like chromosome number, volume of the nucleus, size of the nucleus etc., but on the complicated complex of factors which are far from adequately and completely studied.

We offered the hypothesis that the high resistance of plants to radiation effects is controlled by the complex physiological and biochemical peculiarities of the radioresistant plants themselves.

Specific metabolites were found in the seeds of radioresistant plants. These

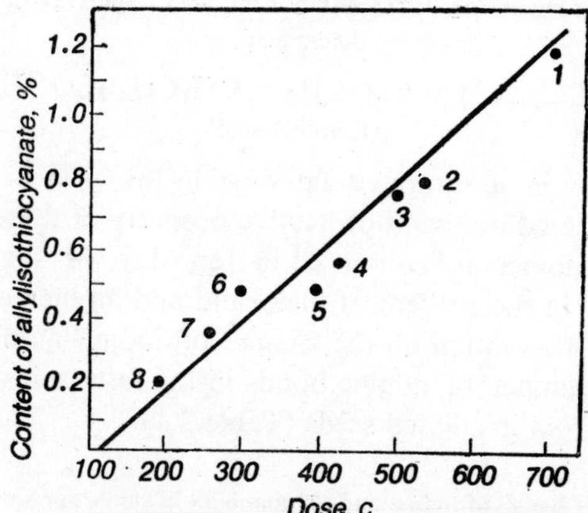

Fig. 2.2. Correlation between the content of allyl isothiocyanate in the seeds of some cruciferous plants and their radiosensitivity: *1*—black mustard; *2*—swede; *3*—Skorospelka (early maturing) variety of mustard; *4*—radish; *5*—garden radish; *6*—late cabbage; *7*—turnip; *8*—early cabbage.

we think can be possibly considered endogenic radioprotectors. In the group of such radioprotectors, we included the allyl isothiocyanate contained in the seeds of cruciferous plants, which are formed from glycoside of synergin due to the action of the enzyme myrosinase on it; cyanide, which is found in the seeds of flax, clover, lucerne and sorghum, is formed from glycosides—linomarin and durrin; unsaturated fatty acids are contained in the seeds of oilseed crops.

These highly reactive substances present in seeds at the moment of irradiation can catch atomic hydrogen during the recombination of free radicals, thus reducing the level of peroxides, hydroperoxides and other highly oxidized compounds formed during irradiation and which can reduce the radiosensitivity of plants raised from the seeds in which they are contained. A direct correlation between the allyl isothiocyanate content in the seeds and the radiosensitivity of different representatives of cruciferous plants is shown in Fig. 2.2.

Further, we observed the behavior of unsaturated fatty acids in irradiated sunflower seeds in order to explain their protective role. It was found that during the irradiation of sunflower seeds their content of unsaturated linoleic acid with two double bonds increases and the content of oleic acid with a single double bond reduces. The transition from unsaturated acid with a single double bond to acid with two double bonds is possible as a result of the removal of two hydrogen atoms from the unsaturated acid with a single double bond [83]:

$$CH_3(CH_2)_4CH_2\text{-}CH_2\text{-}CH_2\text{-}CH = CH(CH_2)_7COOH$$

(Oleic acid)

$$\rightarrow CH_3(CH_2)_4CH = CH\text{-}CH_2 = CH(CH_2)_7COOH + H_2.$$

(Linoleic acid)

The above changes in unsaturated fatty acids take place due to radiation and can be considered one radioprotective property of these acids.

The same phenomenon is observed in four-day-old sunflower seedlings, viz., a reduction in the content of oleic acid and an increase in the content of linoleic acid. Observation on the change of iodine number confirms the increase of the number of double bonds in the fatty acids contained in the seedlings raised from irradiated seeds (Table 2.1).

Table 2.1. Change of iodine and pH numbers in sunflower seedlings with different doses of radiation

Dose, c	Iodine number	Acid number
Control	96 ± 2.8	16.8 ± 0.7
40	100 ± 1.6	13.4 ± 1.6
100	102 ± 0.2	8.1 ± 0.7

We then suggest that the difference in the intensity of respiration which is one of the first and basic physiological processes controlling the activization of life in dormant seeds, can also be the reason for dissimilar radiosensitivities because of intensified repair processes during more intensive respiration. In the comparison of data about the respiration intensity in radioresistant plants of mustard, flax, garden radish and the radiosensitive plants of peas and beans, we observed a more intensive respiration in the plants of the first group.

Our hypothesis shows the correlation of plant radiosensitivity with the biochemical composition of seeds and the physiological processes taking place subsequently in the irradiated plants and controlling the course of further metabolic and repair processes.

At a 1961 conference in Karlsruhe on the effects of ionizing radiation on seeds, numerous reports were presented on the problems of changes in the radiosensitivity of plants due to environmental factors. Particular attention was paid to such factors as humidity, temperature, availability of oxygen and the effects of the conditions and periods of storage of the irradiated seeds.

Most of these research projects were conducted on barley. It is the most convenient object as it is easily subjected to modifying factors, making it possible to reveal the role of moisture corresponding to the effects of irradiation because after hard drying the seeds are easily restored to normal vitality. Kol'dekott dried barley seeds for his experiments to a 4% moisture level in a drying cabinet at 75°C for 24 hours. Wolf experimented with barley seeds with a 2% moisture content. He dried them for 30 minutes at 90°C. Niland, et al. [328] dried barley seeds to 1–3% moisture by keeping the seeds over calcium chloride for not less than two months. N.M. Berezina and R.R. Riza-Zade dried maize seeds to a 4–5% moisture level [23, 24].

All the authors working with seeds containing different moisture percentages observed that ultra-dry seeds at the same radiation doses were damaged 10–12% more severely than seeds with a conditioned moisture content.

In the works of Adams and Niland [260] and Wolf and Sicard [364] it has been shown that during the storage of irradiated barley seeds the radiation injury changes not only depending on the dose but also on storage conditions. The criteria of estimation in their experiments were the seedling height and the chromosome aberrations.

The authors irradiated barley seeds with a dose of 30 c to observe the change of radiosensitivity emerging in seeds with different moisture contents depending on storage conditions. The moisture content of the first specimen was 10% and that of the second, 2%. Seeds with 2% moisture content were found to be more radiosensitive. The height of seedlings grown from these seeds was about 50% the height of the control seedlings. Seeds with 10% moisture content were considerably less radiosensitive and the height of seedlings grown from these seeds was about 80% of the height of the control seedlings. The irradiated seeds were stored for 32 days and storage conditions were not uniform. Depending on the initial moisture content of seeds and the storage conditions, the effect of radiation was sharply different. If the seeds with 10% moisture content were stored under room conditions ensuring the initial moisture level, the depression of seedling growth was maintained at about 80%. Storage of seeds with 2% moisture in the drying cabinet ensured no change in the moisture content of the seeds and height of seedlings raised from them But as soon as the storage conditions changed, i.e., seeds with 2% moisture were kept under room conditions and seeds with 10% moisture in the drying cabinet, the effect of radiation changed drastically: the growth (height) of seedlings grown from seeds with 10% moisture was stunted up to 50% and the damaging effects of radiation which appeared during the irradiation of seeds with 2% moisture were eliminated by the oxygen supply and the duration and storage conditions of the irradiated seeds.

In this intelligently laid out experiment, it was shown how drastically the seedling height changes, depending on the moisture content of seeds at the time of their irradiation and storage. It is also evident from the experiment that some part of the injuries can be restored by changing the storage conditions of the irradiated seeds.

These very authors showed the effect of thermal treatment on irradiated barley seeds. Air-dried seeds irradiated with doses of 20 and 30 c were heated at 90°C for 15 and 30 minutes. The results are shown in Table 2.2.

It is evident from the table that the post-radiation heat treatment of seeds can decrease their radiosensitivity. Results of the experiments are also shown in Fig. 2.4. N.M. Berezina and A.A. Narimanov [36] conducted an analogous experiment on the heat treatment of cotton seeds and showed that the post-radiation heat treatment of seeds for one day after their irradiation at the lethal dose of 40 c restored vitality to plants grown from them. While the cotton seeds without heat treatment died, the treated ones passed through all developmental stages and even produced fruit. However, they drastically lagged behind the control in growth, development and yield. Storing the irradiated cotton seeds for one year also reduced the damaging effect of irradiation, according to the data of these authors [43].

Table 2.2. Post-radiation heat treatment of air-dried barley seeds

Radiation dose, c	Duration of heat treatment, minutes	Seedling height, cm	Seedling height vis-a-vis control, %
Control	Without heat treatment	15.84	—
Control	30	14.09	89.0
30	Without heat treatment	4.39	27.2
30	30	6.77	48.0
20	Without heat treatment	7.60	52.4
20	15	10.40	71.7

According to the data from Nilan, et al. [328], post-radiation treatment of seeds at low temperatures does not change their radiosensitivity but, for the time being, only delays the appearance of the depressing effects of irradiation. For example, in the case of barley seeds irradiated with a dose of 10 c and stored at −80°C and +23°C (Fig. 2.5), the damaging effects of irradiation appeared immediately after irradiation only on seeds stored at −80°C depressing the seedlings whose height was 50% that of the controls. Seedlings from seeds stored at −80°C lagged behind the control by 15% and it was only after the seeds were transferred to a +23°C room before germination that the same growth depression appeared in them as in seeds which were not subjected to post-radiation low temperature treatment.

The degree of injuries due to irradiation changed in seeds with different moisture content and germinated under aerobic and unaerobic conditions.

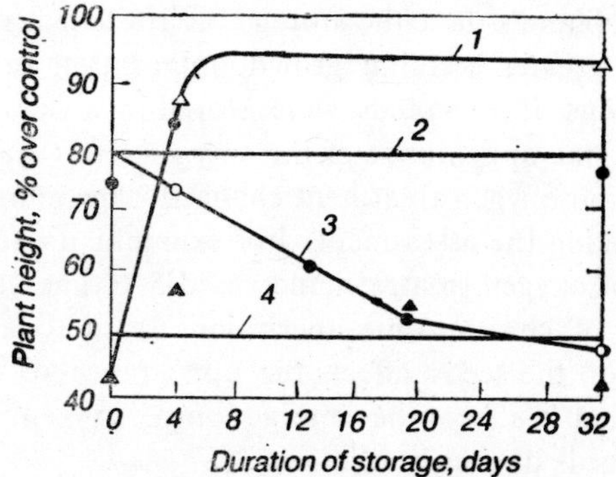

Fig. 2.3. Effect of the conditions of post-radiation storage of seeds on the growth of barley seedlings:

1—seeds with 2% moisture stored under room conditions; *2*—seeds irradiated under room conditions; *3*—seeds irradiated under room conditions and stored over calcium chloride; *4*—seeds with 2% moisture stored over calcium chloride.

Fig. 2.4. Effects of post-radiation treatment of seeds at high temperature (90°C):

a—seedlings grown from the seeds irradiated with a dose of 30 c; *b*—the same as above + heat treatment for 30 minutes.

Maximum oxygen effect in Nilan's experiment [328] was observed in the germination of seeds with 4–5% moisture. The oxygen effect was reduced with an increased moisture content of seeds to 12–16%. This very phenomenon was observed in Kertis' experiments which showed an intensification of post-irradiation injuries in much drier seeds during their germination under aerobic conditions. According to his data, post-irradiation injuries were more distinct on the roots of the seedlings than on the shoots. However, if the seeds were dried before irradiation at high temperatures, i.e., they were subjected to heat treatment or, more correctly, to heat shock, the conditions of aerobic and unaerobic germination did not affect the growth and development of seedlings. In the way, according to the Kertis' data, the heat treatment of seeds before their irradiation has a protective effect.

According to Niland's data the storage of irradiated seeds under aerobic conditions causes greater seedling growth retardation than storage under unaerobic conditions. If the seeds were stored in a vacuum for some days before irradiation, the appearance of the oxygen effect was delayed. Indices of the post-radiation oxygen treatment changed depending on the character considered for making the assessment. For example, the frequency of mutations in the case of oxygen treatment increased 5–6 times, but seedling height and the frequency of chromosome aberrations seven to eight times. As the moisture content of the seeds affects the appearance of the oxygen effect, we can presume that it affects the interaction of oxygen with the free radicals emerging in seeds during irradiation.

Klingmüller [309] studied another factor, the effect of light, on the intensification of radiation injuries of the dried horse-bean seeds stored under aerobic conditions after irradiation.

Randolf and Haber [339] determined the amount of radicals in the embryo, cotyledons and seed coats of lettuce seeds. The radicals were found in all three fractions and their maximum concentration was in the embryo which was about 80% of the total in seed. In dry irradiated seeds the radicals survived for many days.

The life span of radicals depends on the storage conditions of seeds after irradiation, particularly on their moisture content. Radicals decreased during the drying and heating of the irradiated seeds and disappeared when seeds were kept in boiling water.

This research, so conducted, enabled us to presume that temperature, availability of oxygen and moisture, affect plant radiosensitivity through the free radicals emerging in the seeds during irradiation.

While determining the number of paramagnetic centers in the seeds of flax and mustard, the reduction in the quantity of free radicals has been shown during the storage of irradiated seeds [41].

The experiments of Fischnich, Pätzold and Heilinger [280] conducted on potato also confirm the changes of radiation effects depending on the storage conditions and physiological condition of the tubers. The tubers of the potato varieties Erstling, Vera, Corona, Bona, Olympia and Akerzeger were irradiated with doses of 0.005–4 c (Fig. 2.6). The tubers were taken with dormant, underdeveloped and well-developed eyes (buds) removed before irradiation and after it; here the change of storage conditions was studied. Irradiation with the same dose led to different results: the appearance of stimulation, its absence or depression in the development of plants. For example, irradiation with a dose to 100 r on unsprouted tubers of the varieties Bona and Akerzeger produced stimulation. But the irradiation of sprouted tubers of the same varieties led to the depressed growth of shoots. If the shoots were removed from the tubers of these varieties, before irradiation, then the length of redeveloped shoots was indistinguishable from the length of shoots from the control tubers.

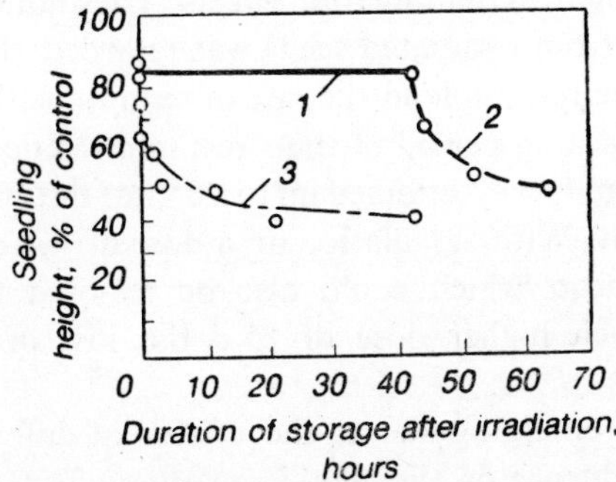

Fig. 2.5. Effect on the seedling height of post-radiation treatment of seeds at low temperatures:

1—with the storage of seeds at −80°C; *2*—after transferring seeds to room conditions which were −80°C; *3*—with the storage of seeds at +23°C.

The temperature at which potatoes were stored before and after irradiation also influenced the effects of radiation. If the sprouted tubers stored at 5–12°C before irradiation were kept at 18°C after irradiation with a dose of 0.1 c, the shoots of the irradiated tubers surpassed the control with respect to their length. But if these same tubers were stored at 4–7°C after irradiation, there was no growth stimulation of the irradiated tuber shoots in comparison with the control tubers stored at the same temperature. In the case of the irradiation of tubers with dormant buds, the varietal differences were clearly expressed, with the effect of radiation obtained through the same dose, particularly, in the irradiation of tubers of Akerzeger variety with doses of 0.005–0.1 c, the number of shoots and their length did not increase in comparison with the control; irradiation of the tubers of the variety Vera in the same doses led to a significant increase in the number of shoots, but tubers of the variety Bona showed depressed shoot lengths in comparison with the control.

In this way, the final effect of irradiation was influenced by varietal peculiarity, physiological condition and the storage conditions of tubers, particularly the temperatures at which they were stored before and after irradiation.

After effects of low doses of radiation on different potato varieties were studied in the subsequent generations. For example, in the third generation of the irradiated tubers of the potato variety Olympia, much better tuber development was noticed than in the control. On the basis of the experiments thus conducted the authors concluded that low doses of radiation have a stimulating effect on the germination of potato tubers if the temperature conditions after irradiation stimulate enzymatic processes in the tubers.

Stein and Richter [355] irradiated lettuce seeds with doses of 3–9 c and studied the role of light in the radiation effects. They noticed that the growth of lettuce seedlings from irradiated seeds was retarded. The growth retardation was found to be reversible in the case of sprouts with light. White light at low intensity for a long period of time restored the development of seedlings to almost normal; they attained up to 90% of the height of the control after light treatment. With irradiation at a dose of 4.8 c, the lettuce seedlings lost phototropism, which could also be restored through light treatment. With the much higher dose of 25 c, the loss of phototropism was almost irreversible.

The differences emerging due to the effects of different types of radiations have great influence on the final irradiation effects, although they are not yet studied well.

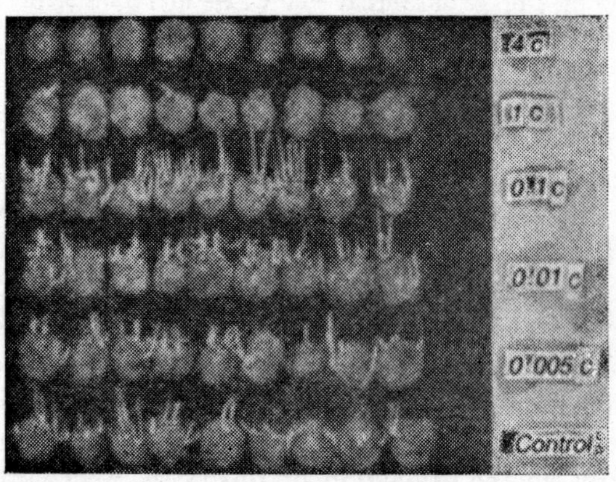

Fig. 2.6. The effect of ionizing irradiation on the germination of the potato variety Vera.

Micke [326] compared the effects of x-radiation at doses of 5, 10, 20, 40, 60, 80 and 100 c with the effects of thermal neutrons used by him in the doses of 2, 5, 10, 20, 40, 80 $\times 10^{12}$ neutrons/cm^2. An integral flux of thermal neutrons at the rate of 5×10^{12} neutrons/cm^2 was approximately equal to the x-radiation dose of 10 c in its effects.

Melilotus alba seedlings died with a dose of 200 c before formation of the true leaves. The survival rate of seedlings with a dose of 60 c corresponded to their survival rate in the integral flux of thermal neutrons at the rate of 80×10^{12} neutrons/cm^2 and it comprised 35% of the control. The author concluded that x-radiation is almost three times more effective than thermal neutrons regarding its biological effect.

V.T. Parfenov [180] irradiated potato tubers with ^{137}Cs gamma-radiation and electrons. Considerable changes in the effects of radiation appeared with different strengths of radiation doses. A.M. Kuzin, N.M. Berezina and O.N. Shlykova [146] compared the effect caused by the same dose of gamma-radiation, but in sharply increased strength. They irradiated air-dried seeds of the corn variety Sterling. The characteristic peculiarity of the radiation effects, in low dosage strength, is the long duration of radiation. Therefore, theoretically it could have been presumed that prolonged irradiation of seeds will have a much stronger effect than acute irradiation at the same dose. In fact, if any biological effect is caused by a brief chain reaction in the micro-structural portions of the object being irradiated and the amount of these portions at the given moment is insignificant but emerge constantly, the effect of acute irradiation will be restricted to a great extent, by the presence of these sensitive portions during irradiation. Prolonged irradiation, as a result of the constantly emerging sensitive portions in the irradiated object can have a great cumulative effect from many short-duration chain reactions. The authors irradiated the first batch of seeds at 0.076 r/min and the second batch with 600 r/min. The cumulative dose of irradiation for both the batches was 11 c. The authors took the length of roots, coleoptile of seed-lings grown from the irradiated seeds of corn and their dry weight as criteria of the radiation effects. As is evident from Fig. 2.7, long-term irradiation had a greater depressing effect than short-term irradiation, particularly at early stages of development.

While irradiating potato with ^{60}Co gamma-radiations V.T. Parfenov and V.S. Serebrenikov observed drastic change of the cumulative dose of radiation depending upon the strength of dose with which irradiation was done [181, 213]. Potato varieties Priekul'skii and Lorkh were irradiated in the doses with strength of 300 and 60 r/min. In case of the strength of dose being 300 r/min the stimulation dose was 300 r, but with the strength of dose of 60 r/min, the optimal stimulation effect appeared in the dose of 150 r. Increase of the strength of dose up to 1,500 r/min led to increase of the optimal stimulation dose up to 1,000 r.

Natar'yan and Marik also observed the changes of radiation effects depending on the strength of the dose while irradiating dry barley seeds with 10 c with strength of the dose being 1 and 0.13 c/hour. The seeds were germinated immediately after irradiation and after storage for 1, 2, 4, 7, 14, 21, 32 days (Fig. 2.8). Differences in the dosage strength were sharply noticed only when the seeds were sown immediately after irradiation. However, in the case of stored irradiated seeds, the differences produced by different dosage strengths reduced slowly with the increased storage period and after 7–32 days of storage, they almost disappeared.

Interesting data [33, 83] showed that dosage effects change in accordance with the time of the year the seeds were irradiated. The essence of these experiments was that mustard, sunflower and castor seeds were irradiated every month in lethal doses for one year and then grown in petri-dishes

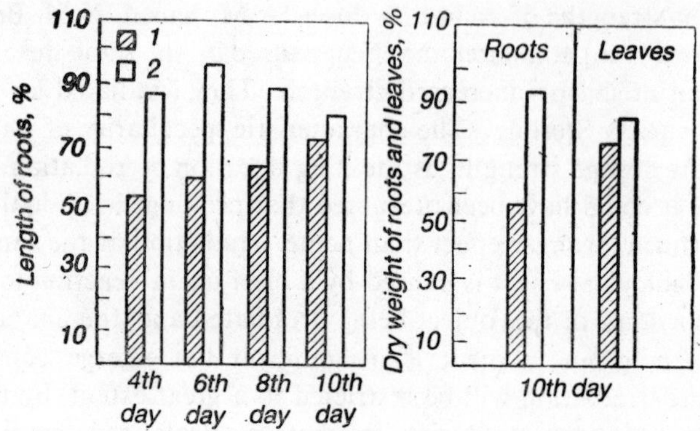

Fig. 2.7. Effects of long- (*1*) and short-term (*2*) irradiation in the dose of 11 c on the growth and development of corn seedlings.

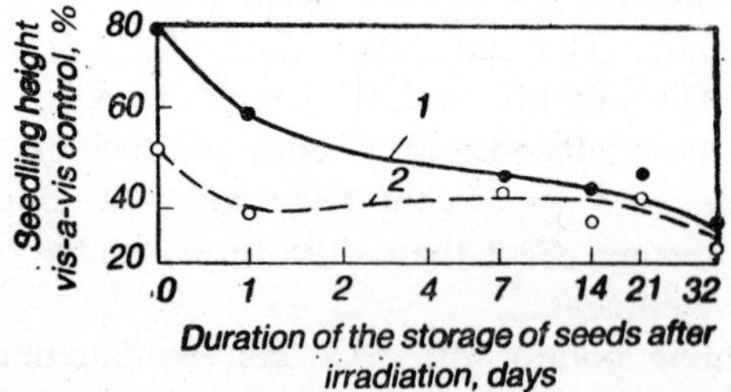

Fig. 2.8. Change of radiation effect during storage of the irradiated seeds depending upon strength of the dose 1 c/hour (*1*) and 0.13 c/hour (*2*).

under constant light intensity and length of the light day (photo period), temperature and humidity in luminostat. The length and aerial part of seedlings were taken as the criteria to evaluate the damaging effects. The experiments showed that during the spring and summer months the effects of lethal doses weaken significantly and the effects of stimulating doses of irradiation are reduced (Fig. 2.9). Therefore, without preliminary tests we should not use the optimal doses in field experiments which are established during winter in the laboratory experiments. Analogous data were obtained by the German researcher Barnetsky [264].

Far from being a complete account of factors changing the effects of irradiation, this testifies to the extreme liability of these effects and their complex correlation, not only with the dose of radiation, but also with the complex environmental and physiological conditions of the object being irradiated.

Presuming that the reason for the poor reproducibility of the stimulation effect in case of presowing irradiation of seeds is the under-assessment of (and at times ignorance about) the role of these above modifying factors.

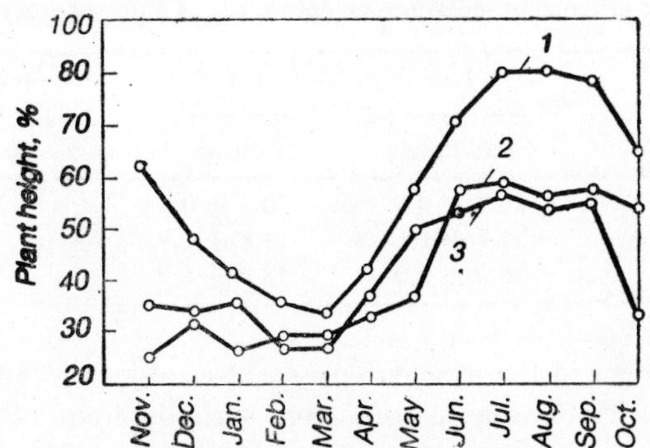

Fig. 2.9. Effect of seasons on the growth of plants from seeds irradiated at the dose of 40 c:
1—castor; *2*—sunflower; *3*—soybean.

We conducted laboratory and field experiments to reveal those conditions which must be observed to obtain stable results. The effect of such factors as the age of the seeds, their moisture content, admissible periods and the storage conditions of the irradiated seeds etc. was studied on corn seeds [35, 23, 24, 26, 27].

In the irradiation of corn variety Sterling seeds stored for many years after harvesting, it was noticed that the seed radiosensitivity changed as they aged and the doses stimulating the growth and development of the seeds from the preceding years harvest may or may not stimulate or may even depress much older seeds [24]. In 1959, while irradiating corn seeds harvested in 1955, 1956 and 1958 at 500 r, the authors obtained the following results (Table 2.3).

The data obtained indicate that when establishing the optimal stimulation dose for seed irradiation it is always essential to consider their age.

In an attempt to study the correlation of the radiobiological effects with the dose, in seeds conditioned at 10–12% moisture content, the corn seeds of variety Sterling were irradiated in the doses of 0.2, 0.3, 0.4, 0.5, 0.8, 1, 2, 4, 10, 20 and 40 c.

The dose curve obtained through measurements of roots on the sixth day after germination is shown in Fig. 2.10. As is evident from the diagram, at 10% conditioned moisture content of seeds the doses of 0.2–0.8 c stimulated the growth of roots. The maximum increase of growth to 25% was observed on the sixth day of vegetative growth. The depressing effect emerged with doses of 1–2 c and at 4 c stimulation recurred, but there was a drastic recurring depression leading to the death of the seedlings with doses of 20 and 40 c. The same double apex nature of the dose curve was observed by K.I. Sukach and E.D. Morozova [219] while irradiating air-dried seeds of the corn variety Moldavanka oranzhevaya.

Table 2.3. Effect of ionizing radiations on corn seeds of different age at a dose of 0.5 c

Year	Length of roots, cm		In comparison with the control, %
	Irradiated	Control	
1958	26.7 ± 0.9	20.3 ± 0.8	131
1956	22.7 ± 1.1	19.8 ± 0.9	114
1955	16.5 ± 2.0	18.0 ± 0.9	91.7

Distinct changes in the dose curve were observed while studying the effect of these doses of irradiation on corn seeds with different moisture content. It is evident from Fig. 2.10 that with seeds dried to 5% moisture content, the optimum stimulation was observed during irradiation with 0.4 c, the depressing effect of irradiation at 0.5–0.8 c, but the recurring stimulation at 1 c, i.e., the radiosensitivity increases with the drying of seeds.

In seeds with 25% moisture content, the radiosensitivity increased still more, optimal stimulation effect is observed with a dose of 0.25 c, but at 0.5 c there occurs a depression. In this way, the change in the conditioned moisture when soaking seeds to 25% or drying them to 5% increases their radiosensitivity considerably and changes the effect of the stimulating dose. Therefore it is essential to know the moisture content of the batch of seeds before undertaking irradiation.

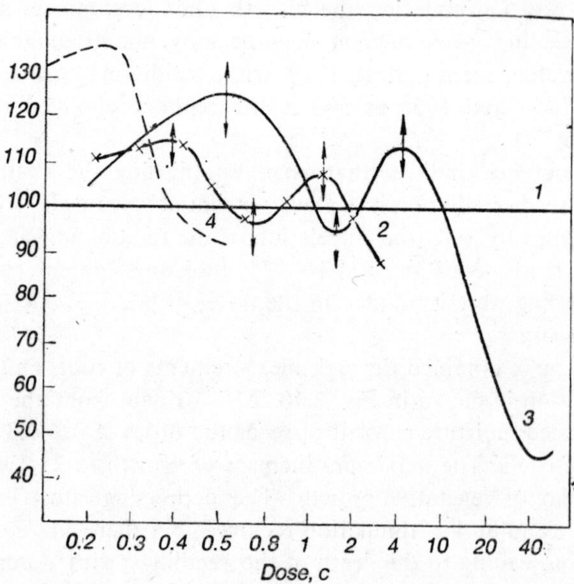

Fig. 2.10. Correlation of the dose curve with the moisture content of corn seeds (along the axis of ordinate—relative length of roots, %):
1—control; *2*—5% moisture; *3*—10% moisture; *4*—25% moisture.

Over a period of several years, while conducting laboratory and field experiments, we noticed that the final effect of stimulation also changes depending on the interval of time between irradiation and sowing. In order to clarify this interrelationship, an experiment was conducted in which corn seeds irradiated at 0.5 c were sown at different intervals of time from irradiation, viz., "direct from rays" (immediately after irradiation) and after 1, 2, 7 and 14 days. Seeds sown immediately after irradiation had up to 70% longer seedlings in comparison with the control (Fig. 2.11). When the irradiated seeds were sown after 1, 2, 7 and 14 days after irradiation, the increase in the length of seedlings was 28–35%. These data indicate that irradiated corn seeds sown immediately after irradiation show maximum stimulation which drastically decreases during the first few days. In field experiments of sowing corn immediately after irradiation, after one month and after one year, the grain yield in the first case (i.e., sowing immediately after irradiation) exceeded the control by 35.7%, by 7% in sowing after one month; but in seeds sown after one year, the stimulation effect was found to be absent (Table 2.4).

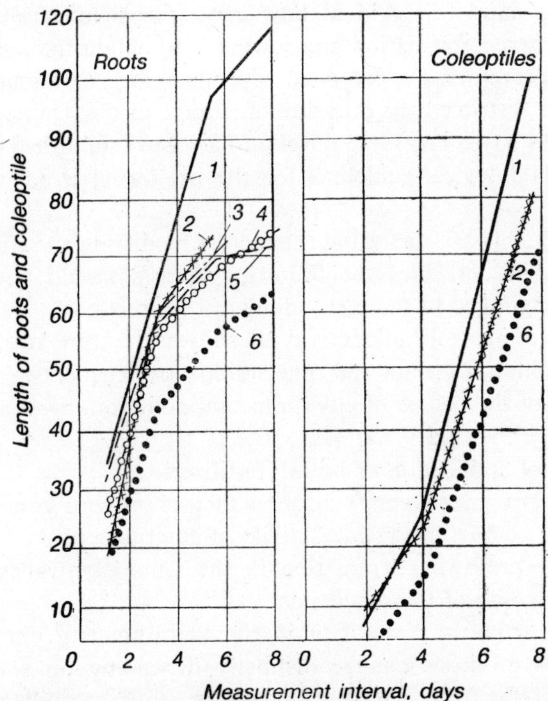

Fig. 2.11. Correlation of stimulation effect with the interval of time between irradiation and sowing:

1—immediately after irradiation; *2*—after one day; *3*—after two days; *4*—after seven days; *5*—after 14 days following irradiation; *6*—control.

Table 2.4. Yield of corn depending on the interval of time between irradiation and sowing for seeds irradiated at 0.5 c

Time of sowing	Green mass		Cobs	
	yield, q/ha	share vis-a-vis control, %	yield, q/ha	share vis-a-vis control, %
Control	310	100	98	100
Immediately after irradiation	407	128	133	135.7
After 30 days	350	112.9	105	107.0
Control	295	100	77	100
After one year	294	99.6	77	100

Analogous results were obtained by N.M. Berezina and M.A. Abdulaev in their experiments [31] with hot house tomato and in the Georgian Institute of Agriculture by F.A. Dedul' [85] who studied the effects of storage on corn and wheat.

From these data it is evident how great the significance of the time interval of between irradiation and sowing is to obtain the recurring stimulation effect. Therefore, to get good reproducibility of stimulation, in all our works we restricted the duration of storage of the irradiated seeds to a maximum of 14 days. The same duration has been established in the instructions and methodological guidelines for the presowing irradiation of seeds [163, 164].

N.G. Zhezhel' [103], analyzing the factors modifying the effect of stimulation doses, considers that the following factors should also be included among those envisaged in the methodological instructions:

—the level of natural radioactivity of soils in the context of the content of uranium, radium, thorium and other radioactive elements;

—the level of the degree of environmental pollution by radioactive fallout, particularly ^{90}Sr and ^{137}Cs;

—the level of application of potash fertilizers.

The author proposes to unify all three factors into one concept, viz., the level of natural and artificial radioactivity of environment.

Data about presowing irradiation in the Chuiskaya valley of KirgSSR confirm the necessity of these addenda.

It is evident from the given data in the literature and our experiments that the effects of doses change distinctly depending on several factors, whose list is still far from complete. Therefore, while compiling methodological instructions on the presowing irradiation of agricultural crop seeds [163, 164] we recommended the following conditions be observed: to irradiate seeds from the preceding year's harvest; to observe strictly the moisture

Table 2.5. Optimum stimulation doses of gamma-radiation established for different agricultural crops

Crop	Variety	Region of trial	Radiation dose, r	Researcher
1	2	3	4	5
Cereals:				
Corn	Sterling	Moscow province	500–1,000	N.M. Berezina, R.R. Riza-Zade
	Hybrid Bukovinskii 3	-do-	500–750	N.M. Berezina, R.R. Riza-Zade
	Hybrid Bukovinskii 3 and 1	Leningrad prov.	500	N.G. Zhezhel'
	Hybrid Bukovinskii 3	-do-	500	N.F. Batygin
	Moldavanka Oranzhevaya	MSSR	500–1,000	K.I. Sukach
	VIR-25	Leningrad prov.	500–1,000	N.F. Batygin
	VIR-42	MSSR	500	V.N. Lysikov, K.I. Sukach
	Sterling	-do-	700	V.N. Lysikov, K.I. Sukach
	Vyatka	Moscow prov.	1,000	L.P. Breslavets
	Viner	-do-	500	O.K. Kedrov-Zikhman
	Viner	Leningrad prov.	1,200	N.F. Batygin
Barley	Nutans-75	KirgSSR	1,000–2,000	N.I. Polyakov, L.A. Sergeeva
	Nakhichevanskii	AzSSR	300	R.R. Riza-Zade
	Viner	Leningrad prov.	1,500–2,000	V.N. Savin
	Pirka and Viner	-do-	500	N.G. Zhezhel'
Wheat	—	Moscow prov.	2,500	O.K. Kedrov-Zikhman
Spring wheat	Diamait, Zarya	Leningrad prov.	500	N.G. Zhezhel'
Oats	Shark	AzSSR	300–500	R.R. Riza-Zade
	Zolotoi dozhd'	Leningrad prov.	500	N.G. Zhezhel'

(*Contd.*)

1	2	3	4	5
Legume pulses:				
Peas	Spartanets, Grei Nikai	Leningrad prov.	250	N.G. Zhezhel'
	Kapital	Moscow prov.	350	A.I. Atabekova, L.P. Breslavets
	Tergsdag	LatvSSR	300–1,000	V.T. Ore
	Romonskii	MSSR	800–1,200	V.N. Lysikov, K.I. Sukach, V.S. Serebrenikov, D.N. Goncharenko
	Musluchnyi	Leningrad prov.	600	N.F. Batygin
Spring vetch:				
for slage	300–17	MSSR	1,500	V.N. Lysikov
for grain	300–17	-do-	500	S.N. Bloshko
Vegetables:				
Potato	Priekul'skii	Moscow prov.	150–300	A.I. Grechushnikov, V.S. Serebrenikov
	Berlikhengen	-do-	500	N.M. Berezina, R.R. Riza-Zade
	Severnaya roza	-do-	150	V.S. Serebrenikov
	Sovkhoznyi	-do-	300	V.S. Serebrenikov
	Lorkh	-do-	300–500	A.I. Grechushnikov, V.S. Serebrenikov
	Priekul'skii	Leningrad prov.	300–500	N.F. Batygin
Carrot	Nesravnennaya	Moscow prov.	2,000	N.M. Berezina, R.R. Riza-Zade
	Nantskaya	Krasnodar territory	2,500	N.M. Berezina, T.P. Zhadanova

	Shantane	Moscow prov.	2,000	N.M. Berezina, G.I. Shchibrya, M.L. Romanchikova
Cabbage	Kolkhoznitsa	Moscow prov.	2,000	N.I. Popova
	No. 1	-do-	2,000	N.M. Berezina, R.R. Riza-Zade
	No. 1 and Ditmarskaya	Krasnodar territory	2,000–16,000	G.I. Satalkina
Tomato	Teplichnyi	Gor'ky prov.	500–1,000	I.F. Alekzandrova
	Luchshii iz vsekh	Moscow prov.	2,000	M.A. Abdulaev
Cucumber	Klinskii teplichnye	-do-	300	N.M. Berezina
	Teplichnye	Leningrad prov.	500	N.G. Zhezhel'
Dill	—	Moscow prov.	1,000	M.A. Abdulaev
		AzSSR	1,000	T.V. Brazhnikova
Lettuce	—	Moscow prov.	300	M.A. Abdulaev
Garden radish	Rosy with white tip (in the open field)	-do-	1,000	N.M. Berezina
	Saks (not in beds)	-do-	1,000	N.M. Berezina, G.I. Shchibrya
Watermelon	Kakhetinskie	AzSSR	10,000	A.I. Khudadatov
Muskmelon	Kolkhoznitsa	-do-	4,000	A.I. Khudadatov
	Kusorchaiskaya	-do-	2,000	R.R. Riza-Zade, I.G. Suleimanova
Industrial crops:				
Sugarbeet	—	Leningrad prov.	500–1,000	N.G. Zhezhel'
	M–70	LatvSSR	2,000	A.T. Miller
	M–80	-do-	2,000	A.T. Miller
	Verkhnyachka	Leningrad prov.	500–5,000	N.F. Batygin
	Belotserkovskaya	-do-	2,000–6,000	N.F. Batygin
	—	Ul'yanov prov.	2,000–2,500	V.I. Kostin
Sunflower	Peredovik	Krasnodar territory	2,000	T.E. Guseva
	VNIIMK-1646	MSSR	2,000–2,500	V.N. Lysikov
	Armavirskii	AzSSR	2,000	R.R. Riza-Zade

(*Contd.*)

1	2	3	4	5
Peanut	Adyg	Krasnodar territory	750	T.E. Guseva
Castor	VNIIMK-165	-do-	4,000	T.E. Guseva
Flax	I-7	Moscow prov.	1,000	A. Nigmanov
Cotton	108 F	UzSSR	500–3,000	Sh.I. Ibragimov, R.I. Koval'chuk
Ambari-hemp	—	-do-	1,000–5,000	U.A. Arifov, G.A. Klein
Jute	—	-do-	1,000–5,000	U.A. Arifov
Coriander	—	AzSSR	1,000	Zh. Brazhnikova
Soybean	Dnepropetrovskaya	MSSR	800–1,200	V.N. Lysikov, K.I. Sukach
Tobacco	Trapezund	MSSR	1,500–2,500	V.N. Lysikov, K.I. Sukach
Fodder crops:				
Clover	—	Moscow prov.	500–1,000	O.K. Kedrov-Zikhman
	Aleksandriiskii	AzSSR	5,000	R.R. Riza-Zade
	—	Moscow prov.	1,000	A. Nigmanov
Vetch	L'govskaya	Leningrad prov.	250	N.G. Zhezhel'
Lucerne	AzNIKhI-261	AzSSR	5,000	R.R. Riza-Zade
	—	UzSSR	10,000	V.L. Mukhanova
	—	Moscow prov.	1,000	A. Nigmanov
Lupin	Zheltyi Kormovoi	Leningrad prov.	500	N.G. Zhezhel'
	Bezalkoloidnyi	Moscow prov.	4,000–16,000	N.M. Berezina, R.R. Riza-Zade
Turnip	Osterzundomskii	Leningrad prov.	1,000	N.G. Zhezhel'

content of seeds and irradiate only those batches which have conditioned moisture; to observe the optimal storage durations of the irradiated seeds, without letting these intervals exceed two weeks with the exception of those seeds which require longer periods of rest before germination; to irradiate the released varieties of agricultural crops for which preliminarily optimal stimulation doses have been established (for the particular zone).

As a result of many years of research by many researchers in different institutions of various ministries, a list of optimal stimulation doses has been compiled for all major agricultural crops (Table 2.5).

It must be mentioned that different researchers working independently of each other and often with no link to each other obtained very close values of optimal doses for the same crops, noticed the same biological peculiarities of the irradiated seeds, like the enhancement of ripening, change of quality, additional branching, etc., which increases our confidence in the correctness of the observed phenomena.

3. Theoretical Basis of Presowing Irradiation of Seeds

Presowing irradiation of seeds is a new agronomic measure to enhance ripening, increase the yield potential of agricultural crops and improve the quality of the produce. Along with a generalization of data from different institutions and countries about the practical significance of this measure we consider it necessary to shed light on its theoretical base as well. Without understanding the action mechanism of the stimulating doses of irradiation, this measure cannot be implemented and put to correct practical use. While reviewing the developments on presowing irradiation of seeds from a historical aspect, we noticed that there was a desire to give a theoretical explanation of the action mechanism of low doses of irradiation by several radiobiologists who tried to clarify the biological action of the penetration of radiations discovered by Röentgen. However, all these efforts were far from the real understanding of the action mechanism of ionizing radiations whose certain links have no comprehensive explanations even now.

N.V. Timofeev-Resovskii and associates [225–228] put forward a hypothesis to explain the stimulating effect of ionizing irradiation, according to which toxic substances develop in seeds during irradiation.

Under low gamma-irradiation doses, there emerges a light intoxication in the organism being irradiated which increases the general tone of metabolism and leads to increased rates of cell divisions in the presence of a very insignificant amount of abnormal mitosis. The author banks on the theory commonly accepted in pharmacology and toxicology, that low doses of toxic substances have a stimulating effect on living organisms, while high concentrations of these same substances can be lethal. However, the hypothesis advanced by N.V. Timofeev-Resovskii was purely speculative in nature. He did not point out the definite toxic substances emerging during irradiation.

G.R. Rik [203] was inclined to consider the stimulation as a result of a defensive reaction of plants to the action of any irritant, including the ionizing radiations. Primary changes emerging in plants at the moment of irradiation cause numerous secondary changes to which the plants react with a complex, physiological reaction. In the process of vegetative growth, depending on the irradiation dose, these changes exert a depressing or stimulating effect.

N.F. Batygin [20] explains the initial reaction of living organism to the action of ionizing irradiations as a change in the physical and chemical properties of protoplasm which results in the intensification or depression of some physiological processes and metabolic reactions associated with them. As different plant tissues and organs have different sensitivity to the ionizing irradiations, physiological changes of different levels emerge which lead to a physiological qualitative variation in the organ. As a result of interaction with environmental conditions, such qualitative variations can be the reasons for the stimulation or depression of the development of the plant.

Under normal conditions during plant hybridization, the qualitative variation emerges in the hybrid organism as a result of the union of two gametes of different quality. The qualitative variation of gametes in its turn is determined by the level of metabolic processes inherent to parental forms, their degree of affinity and their morphological condition.

When the qualitative variation emerging during hybridization or irradiation drastically exceeds the normal limits, the organism dies; with favorable combinations of these changes there emerges a heterosis or stimulation effect.

V.S. Andreev [8] considers that there is no need to search for special mechanisms for damaging or stimulating doses and he pays attention to the link of biological effects of irradiation with the effects of irradiations on the neucleoproteins of the nucleus; in such case, depending on the degree and nature of injuries, the irradiation can depress or stimulate the growth and development of plants. The author considers the presence of after effects observed in the progeny of the stimulated plants as convincing proof of the correctness of his theory. This phenomenon was noticed by L.P. Breslavets in rye when the seed progeny of the stimulated plants gave a 36% increase of yield in comparison with the control.

N.A. Poryadkova [187, 188] also noticed an analogous phenomenon in the progeny of wheat, barley and peas. While irradiating strawberry suckers over a period of three years N.M. Berezina and others [40] observed the retention of the stimulation effect. While irradiating potato tubers, Fischnich et al. [280] and V.S. Serebrenikov [214] also noticed an increase of yield potential of second and third generation rootcrops. These facts testifying to the appearance of stimulation in subsequent generations indicate the involvement of a genetic factor in the phenomenon of stimulation.

However, the radiobiological effect is not confined to changes emerging during irradiation or immediately after it. In the process of the growth and development of plants raised from irradiated seeds, beginning from seedling emergence and ending with the ripening, there appear new, quite diverse changes manifested in the acceleration of the cell division rate, enhancement of growth and development, change of organogenesis, yield increase and its quality change, i.e., there emerges a very complicated sequence of changes which has been termed the effect of distant irradiation action.

A.M. Kuzin [143–145] gives a more comprehensive idea about the action mechanism of stimulating doses of irradiation and the theoretical basis for the presowing irradiation of seeds. He considers the effect of low doses as a result of biophysical and biochemical advances leading to non-specific and specific gene activation and the depression of individual parts of the genome due to action of different trigger-effectors (different release mechanisms).

For example, to break dormancy and enhance the germination of irradiated seeds, great importance is attributed to low molecular polyphenols and quinones performing the role of non-specific trigger-effectors. To desuppress genes, these substances induce the activation and early synthesis of enzymes that are required to break dormancy and germinate seeds. Thus was shown the radiation activation of tyrosinase, polyphenoloxidase, lipoxidase which led to the increased and early accumulation of orthoquinones and lipid peroxides in the irradiated seeds. These compounds in low concentrations had a stimulating effect on the germinating seeds which made it possible to consider them non-specific trigger-effectors activated by the gene.

Gene activation in the cells of the aleurone layer responsible for the synthesis of α-amylases and proteases takes place under the influence of a specific trigger-effector, viz., gibberellic acid being synthesized in seed embryos during the first few hours they are soaked. The primary acceleration of oxidation processes in the irradiated seeds enhanced the synthesis of gibberellic acid. This intensified the synthesis of amylase and protease and the result was the early and intensive flow of sugars and aminoacids toward the embryo of the irradiated seeds and the consequent enhancement of seed germination and much faster development and growth of the seedlings.

Another example can be the increased tryptophanesynthetase activity in six-day-old corn seedlings grown from the seeds that were irradiated at 500 r. The activation of this enzyme increased the content of β-indoleacetic acid by 26% in the seedlings, which, being an important growth harmone, stimulated the growth and development of plants from the irradiated seeds.

Further development of the irradiated plants takes place with the formation of new effectors desuppressing the genes which are responsible for the synthesis of enzymes and are essential to the plants for subsequent development, e.g., the synthesis of flowering hormones that cause early flowering and ripening in the irradiated plants and the more profuse formation of generative organs.

Thus, the process of stimulation is the result of step-by-step action of non-specific and specific trigger-effectors emerging in the plants at different stages of their development.

Along with the above mechanisms, the stimulation effect can cause processes like the intensified respiration that create an energetic base for stimulation; heterogeneity emergence as a result of different radiosensitivities of individual tissues and enzymes in the irradiated seeds; the removal of apical domination so that, instead of one apical growth point, there develop lateral growth points which causes profuse additional branching to increase the growth of green mass of the plants raised from irradiated seeds.

Now we shall examine individual manifestations of the effect of distant irradiation action which stimulates the growth and development of plants raised from irradiated seeds. One manifestation of the effect is the increased mitotic index in the meristem of seedling radicles. References to the increased mitotic index due to the action of low doses of irradiation are first found in the works of Geiger and Köernike. In subsequent works [16, 54, 149, 227, 236] we find more detailed quantitative material confirming the increased mitotic activity in seedlings from irradiated seeds.

L.P. Breslavets was one of the first researchers to irradiate soaked rye seeds at different doses (0.25–0.8 c). With an increased dose, the amount of mitosis increased, which was at its maximum with a dose of 0.75 c, but it reduced with further dosage increases. The calculations have shown that the increase of the diameter and length of radicles, observed by L.P. Breslavets, due to the action of the stimulating x-radiation dose was in fact caused not by the increased cell size but by their increased number and, therefore, was the result of the increased mitotic activity.

According to data from L.I. Tsarapkin, in plants where the stimulation effect of irradiation was manifest, the average cell size was less but their number and amount of mitosis was more than in the control. With high dose irradiation of plants, the cells were bigger but the frequency of mitosis and number of cells were less. Analogous results have been obtained by E.N. Sokurova [216] on bacterial cultures working with nitrogen fixing bacteria.

In this way, according to the opinions of a number of authors, one of the first reasons for the stimulation of plant growth and development with low irradiation doses is attributed to the stimulation of cell division which is accelerated by the division of the nucleus and, in all probability, with the acceleration of DNA synthesis.

Another manifestation of the effect of distant irradiation action is the change in the organogenesis of plants expressed in the additional branching of vegetative and generative organs of the plants. This phenomenon was noticed with the irradiation of different, taxonomically distant groups of annual and perennial plants. Also with high doses of irradiation, we often noticed the emergence of additional lateral buds as a result of the suppression of the growth of the main or apical bud. With stimulation doses different phenomena appear: the main bud develops completely normally, but along with it begins the growth of additional lateral buds. D.M. Grodzinskii considers this phenomenon as a specific effect of the stimulation doses of radiation due to which the apical domination, which is inherent to almost all higher plants, is removed; in such a case the lateral buds begin to develop simultaneously with the apical bud. One reason for the simultaneous activization of apical and lateral buds, according to N.M. Berezina and A.F. Revin [37], can be the increased quantity of growth substances due to the effect of the stimulation doses of radiation. A much higher content of β-indoleacetic acid, exceeding the control by 18–20%, was observed in corn seedlings as a result of the more intensive synthesis of some predecessors (serine, tryptophane and indole) of this growth hormone.

N.M. Berezina, E.I. Korneva and R.R. Riza-Zade [35], K.K. Roze and V.T. Kietse [206] observed a multicob habit in corn as a result of irradiation. The resultant phenomenon was that a cob developed in almost every leaf axis of the plant and which remained underdeveloped and immature to the end of the vegetative growth of the plant. As many as ten cobs were found in some plants grown in areas of optimal radiation in the gamma-field. For the unirradiated plants, e.g., the corn variety Sterling, the plant typically formed two cobs. In individual cases, groups of additional cobs formed as whole bunches with a profuse number of corn shucks. The main cob completing the so-formed bunch was usually developed thoroughly and it ripened to the milky or milky-dough stage, but the remaining cobs remained underdeveloped at harvesting. The manifestation of this phenomenon was particularly clear in 1961 on corn crops under favorable climatic conditions on the state farm "Zaokskii" with fertile flood-plane soils of the River Oki.

Additional branching was observed in other crops as well. Chronic irradiation of the buckwheat variety Bogatyr' [58] with the total radiation dose of about 0.34 c in a gamma-field with dense planting caused more intensive branching than in the control (Fig. 3.1). The total height of buckwheat plants in this zone was less than in the control, but there were more additional branches and they developed much more vigorously than in the control, frequently exceeding it with regard to the length of the main stem. The average

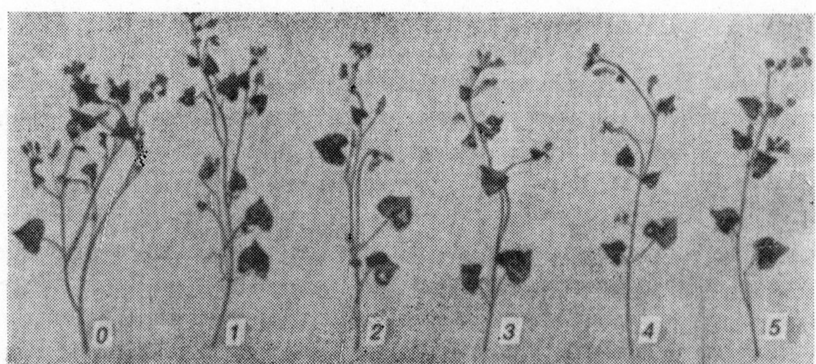

Fig. 3.1. Additional branching of buckwheat due to chronic irradiation on gamma-field in different zones with the total radiation dose (c) of: *0*—0.34; *1*—0.14; *2*—0.2; *3*—0.008; *4*—0.004; *5*—control.

length of the lateral branches in the irradiated buckwheat were 22.6±0.89 cm, whereas it was 7.9 ± 0.8 cm in the control plants. Additional buckwheat branching took place also as a result of the emergence of additional buds due to radiation without suppressing the apical bud.

With the irradiation of ambari-hemp seeds [10] in the doses of 25 and 50 pex (pex—physical equivalent of x-rays, the dose of any ionizing radiation in which the energy absorbed on ionization of the air is equivalent to the dose of 1 r of gamma radiation; this unit is not used now), the main stem branched at a height of 10 cm from the root-collar with the formation of two or three stems, while the control plants always developed only one stem.

Dichotomous branching of clover on irradiation was noticed by Micke [326]. Bifurcation of the main clover stem took place in 5.9% of the stems in the case of seed irradiation at 40 c with x-rays. Clover branching emerged in 3% of the plants due to the thermal neutrons at 40 c.

When cotton seeds were irradiated at 3–6 c, the researchers noticed the appearance of dichotomous stem branching and the formation of bunches of capsules along the axis of one leaf, there were four to five capsules in one bunch.

Irradiation of the organs of vegetative propagation caused analogous morphological changes. When tubers of potato variety Berlikhengen were irradiated with 0.5 c, which exerted a stimulating effect on the growth, development and yield of the tubers, there were changes in the organogenesis: although one shoot developed from each activated bud of the control plants, the shoots of the irrdiated tubers branched off at 1–2 cm from the tuber (Fig. 3.2). The formation of several shoots, instead of one, facilitated the development of more green mass and more intensive photosynthesis increasing the yield potential.

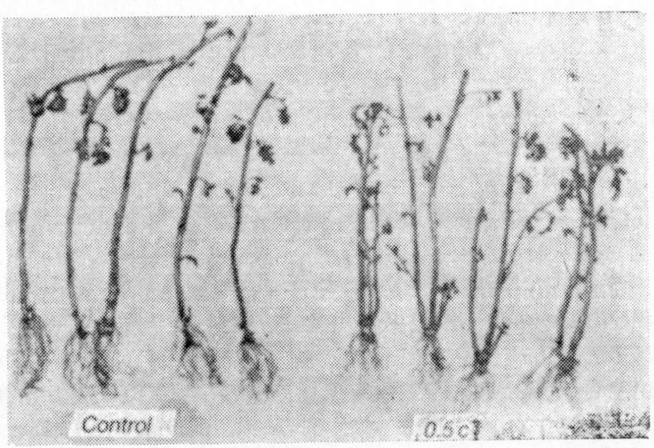

Fig. 3.2. Additional branching of potato shoots due to the presowing irradiation of tubers at 0.5 c.

Irradiating rhizomes of the peppermint variety Prilukskaya with 1 c stimulated the growth and development of plants, evoking a very large number of buds in the irradiated rhizomes as observed in Fig. 3.3. On the average ten shoots were obtained from 8 cm long irradiated rhizomes and only four from the unirradiated rhizomes.

Interesting organogenesis changes were noticed in the experiments conducted on the irradiation of fruit trees grown on the gamma-field of the I.V. Michurin Central Genetic Laboratory. The effect of chronic irradiation was studied on such varieties of apple as Antonovka, Pepin shafrannyi, Rozovoe prevoskhodnoe, Doch' Flavy, Zolotaya Osen and the varieties developed by Chernenko and Tikhonova, as well as on groups of varieties Bere zimnavaya, Michurinka and Bere zheltaya.

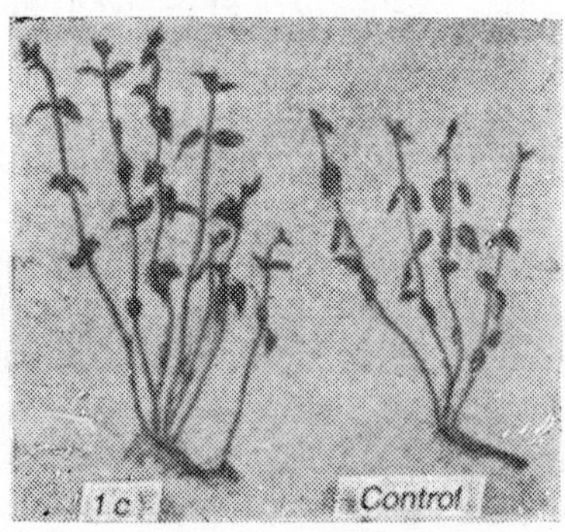

Fig. 3.3. Evocation of additional buds on peppermint rhizomes with presowing gamma-irradiation.

The trees were planted in four zones where, depending on the distance from the source, they received different total dose of irradiation. With a total dose of 178–487 r the saplings showed a tendency for quicker setting up for the time of fruit bearing, but in case of the total dose of 2,450 r these processes were considerably more poorly manifested than in the control plants. Besides, in the irradiated saplings the researchers noticed morphological changes (Fig. 3.4), pseudodichotomous branching, double bifurcation of shoots and the development of four well-differentiated, buds from the apical shoot (Fig. 3.5) in individual cases. Quite often the apical buds did not form at all, but the shoots terminated with a leaf peduncle which carried normal leaf blade. Considerable deviations from the normal were observed in the development of axillary buds: the buds did not form at all in the axis of some leaves but up to three buds were formed in the axis of others. In some saplings considerably more leaves per unit length of shoot were noticed than in the control. Analogous phenomena were observed by the researchers of

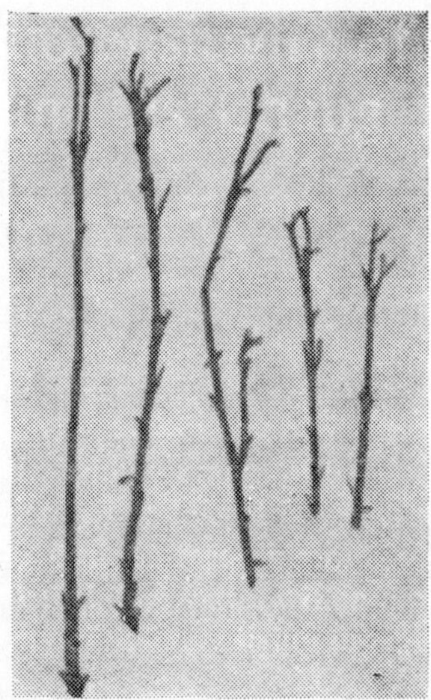

Fig. 3.4. Evocation of additional buds on fruit plants with chronic irradiation.

Fig. 3.5. Evocation of additional leaf buds due to the effects of chronic irradiation.

the Institute of Horticulture in GDR and by G.N. Imamaliev at the Institute of Biophysics of the USSR Academy of Sciences. The experiments of L.P. Breslavets on the irradiation of ambari-hemp also induced dichotomous branching, whereas the plant always produces one stem under normal conditions [53].

In this way, plants belonging to different and quite distant taxonomic groups showed morphological changes due to ionizing radiations. These changes are general; with irradiation the deep-seated buds awaken and as a result the additional branching takes place, resulting in a greater accumulation of vegetative mass or generative organs.

These morphological changes lead to an acceleration in ripening, the increased yields of several crops and the improvement of the quality of produce raised from irradiated seeds. Well planned experiments [276] with the chimera pear variety Red Bartlett showed that radiation facilitates the awakening of deep seated, usually dormant buds. The epidermal layers of the pear tree contain anthocyanins which impart a red color to tree bark.

Through irradiation, the author intended to awaken the deep-seated buds which, on account of being in the tissue beyond the zone of anthocyanin coloration, should have produced green shoots, in his opinion. This theoretical presumption had been confirmed by the appearance of green shoots in the Red Bartlett pear variety after irradiation.

Irradiation of chrysanthemum seeds caused an increase of ligulate flowers and inflorescene and facilitated the formation of more beautiful and luxuriously developed inflorescences, which also occurred as a result of the evocation of additional flower buds.

A.I. Khudadatov [239–241] showed that when seeds of melons and cucumber were irradiated, additional buds are activated on the main stem to form lateral branches. In muskmelon plants grown from irradiated seeds, there were an average of nine stolons on one plant, whereas one control plant had only three or four stolons (Fig. 3.6). One irradiated cucumber plant had six stolons, while the control plant had four or five (Fig. 3.7).

It is worth paying attention to the phenomenon of "relative" stimulation—the cases so termed by us when, due to the effect of radiation, there is a growth and development of organs which have no practical value in the cultivation of this or that crop.

Relative stimulation was obtained in an experiment on the presowing irradiation of castor seeds with doses of 4–8 c. The development of lateral shoots was sharply stimulated due to effect of this irradiation. There was up to a 60% increase in the weight of green mass in comparison with the control. However, the yield of seeds, for which this crop is grown, was adversely affected due to the increased green mass. Capsules in the additional clusters did not mature completely and failed to produce seeds, but in the process of their development they weakened the main cluster and reduced its productivity.

One more biological peculiarity of irradiated plants is well known, viz., enhanced ripening. The phenomenon of enhanced ripening was noticed in almost every experiment beginning from those of Köernike and Geiger to modern Russian and foreign researchers. According to the data from A.V. Doroshenko [93], whose objective was to study this particular effect of radiation, enhanccd ripening by 7–14 days was noticed in the irradiation of the seeds of oats, millets and rapeseeds. During this period the irradiated

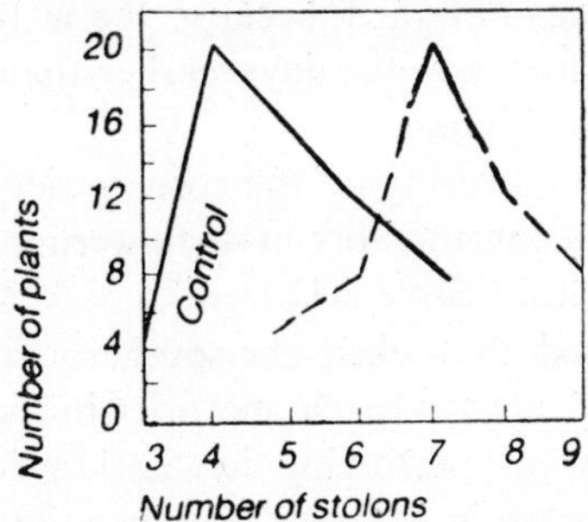

Fig. 3.6. Formation of additional stolons in muskmelon as a result of presowing irradiation of seeds.

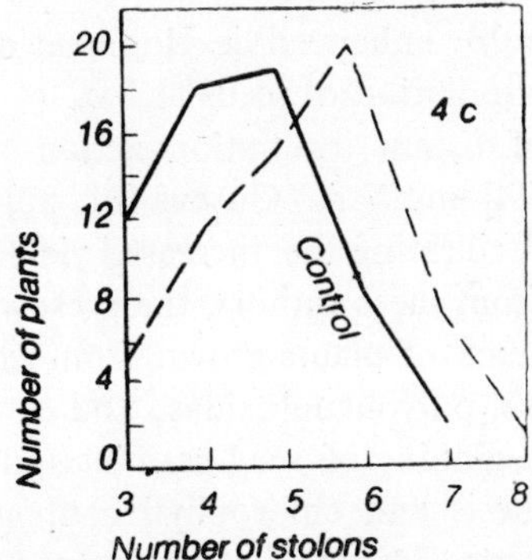

Fig. 3.7. Formation of additional stolons in cucumber as a result of presowing irradiation of seeds.

plants attained great height and better development which facilitated a yield increase. As per the data from L.P. Breslavets, et al., irradiating radish seeds enhanced the maturity of the tubers by four to five days. Estimates made by N.M. Berezina regarding the yield of the hot house radish grown from irradiated seeds showed that the first harvest from an equal area yielded 194 bundles (of ten tubers each) of the control radish and 310 bundles of the irradiated radish, i.e., almost 60% more. This yield increase in the first harvest was the result of the fact that tubers raised from the irradiated seeds matured much earlier. In subsequent harvests the increase in the yield of irradiated radish gradually reduced until it was the level of the control yield in the last harvests. The total yield increase of the irradiated tubers in this experiment was 26%.

When strawberry was irradiated in the gamma-field, 80% more berries were obtained in the first harvest in comparison with the control. The last harvests gave indices almost similar to those of the control.

We observed an analogous picture of ripening with the irradiation of cucurbitaceous plants. In these crops the maximum economic effect from presowing irradiation was obtained because the watermelons, muskmelons and cucumber matured five to seven days earlier than the control, when the price of this produce was highest.

Observations on the flowering of the corn variety Sterling showed that at first counts there were considerably more flowering plants among irradiated ones than in the control (Table 3.1).

It must be mentioned that when the southern crops are moved northward where they do not always reach maturity in the new regions of cultivation, the effectiveness of presowing irradiation increases considerably. With the cultivation of corn in Moscow province, its enhancement in development as a result of presowing irradiation is manifested not only by greater green mass, but also in much earlier maturity of cobs which increases the fodder value of the green mass of the corn.

We noticed considerably enhanced development of cabbage heads of the variety Slava with the irradiation of seeds at 2 c.

Among the effects of distant irradiation action on plants, P.A. Vlasyuk and associates [65, 66, 74] and V.A. Guseva [81, 82] point out the multiple physiological changes facilitating the increased yield and improved quality. According to the data from these authors, the presowing irradiation of seeds affects the enzyme systems of plants grown from irradiated seeds, reduces the activity of peroxidase, polyphenoloxidase and catalase, which was somewhat increased in the beginning of seed germination and also increases the intensity of photosynthesis and chlorophyll content. The authors explain that this phenomenon is associated with the increased photochemical activity of chloroplast due to the irradiation. In addition to all these phenomena, each of which can be a stimulus to increase the yield potential of plants, the authors have established that because of the effects of irradiation there is an intensified absorption of microelements by the roots of irradiated plants. When ashes from the irradiated plants were analyzed they were found to have large contents of such microelements as titanium, cobalt, beryllium, nickel, silver and others. More intensive absorption of microelements from the soil with the presowing irradiation of seeds is considered by the authors as one factor which facilitates the increased yield potential of the plants. However, the final effect of irradiation depends also on which plant species was subjected to irradiation and what type of metabolism is characteristic to that particular plant species. If one plant species during its evolution has developed biochemical systems for quickly transforming carbohydrates and fats and other species a system to transform carbohydrates into cellular fibers, then, due to influence of the complex effects of the distant action of presowing irradiation, the oil content in the first plant species increases and the quantity of fibers increases in the second species.

These important problems pertaining to the mechanisms of initial and distant irradiation effects with the presowing irradiation are being studied at present.

Table 3.1. Number of flowering phases in corn with the presowing irradiation of seeds with gamma-rays ^{137}Cs with the dose 500 r

Year	Treatment	No. of plants counted	Without tassels		Tassel emergence		Tassels formed		Flowering plants	
			Number	Percentage of the total	Number	Percentage of the total	Number	Percentage of the total	Number	Percentage of the total
1961	Control	200	60	30	85	42.5	50	25	5	2.5
	Irradiated	292	31	10.6	120	41.1	110	37.6	31	10.6
1962	Control	200	50	25	54	27	53	26.5	43	21.5
	Irradiated	200	17	8.5	53	26.5	67	33.5	63	31.8
1963	Control	200	93	46.5	52	26.0	24	12.5	31	15.5
	Irradiated	200	92	26.0	38	19.0	29	14.5	41	20.5

4. Presowing Irradiation of Seeds: Results of Experiments

VEGETABLES AND CUCURBITS

Seeds of vegetable crops were the object of much radiobiological research. The most extensive experimental material has been gathered on these crops.

The first conference on the presowing irradiation of seeds [1961] recommended that work be expanded in this direction, particularly on vegetable crops because it opens the prospects of increased yields of costly produce and high profits which would quickly repay the expenditure on irradiation. Seeds of most of vegetable crops are small and are sown in small quantity which also makes irradiation cheaper. Finally, presowing irradiation, as will be seen from the material presented below, almost always accelerates the ripening of vegetable crops and increases their vitamin contents, which is particularly important in the early spring period.

Radish. It is one of the crops investigated by several researchers. A biological peculiarity of this crop is its very high radiosensitivity which was noticed in the initial studies by Köernike [303] who conducted experiments on the irradiation of radish. The seeds of different radish varieties like Saks, Rozovyi with white tip and Rubin were irradiated but no significant varietal differences in radiosensitivity could be established.

L.P. Breslavets, et al. [58], studied the effect of x-radiation with 0.25–2 c doses with the strength of 425 r/min on radish seeds under the voltage of 188 kV and 6 mA current strength at 60 cm from the x-ray tube. Irradiation was done with a cardboard filter. They used first class seeds of the variety Saks with 12% moisture content. Under laboratory conditions when seeds were germinated in Petri dishes there was a 3–22 mm increase in the root with 0.25–1 c doses; the optimal dose was 0.5 c.

Next year the seeds irradiated with 0.5–1 c were sown in one meter plots in three replications (Table 4.1, Fig. 4.1). As may be seen from the given data, the field experiments confirmed the stimulating effect of the 0.5–1 c doses. The maximum increase of tuber yield (by 33%) was obtained through irradiation at 1 c, i.e., with a much higher stimulating dose than the optimal dose used under laboratory conditions.

In subsequent experiments seeds of the radish variety Rozovyi with white tip were irradiated with ^{60}Co gamma-rays. The optimal stimulation dose was found to be 1 c, i.e., the same as the x-radiation for these seeds.

Table 4.1. Yield of tubers of the irradiated radish in field experiments (average of three replications)

Dose, c	Yield from the plot, g	Yield in comparison with the control, %
Control	3,744	100.0
0.5	4,485	119.7
1	4,987	133.4

Fig. 4.1. Effect of the presowing irradiation of seeds of the radish variety Rozovyi with white tip.

Gamma-irradiation of seeds at 1 c during a period of three years under field conditions helped produce an additional 11 to 30% of tubers. In 1958 commercial crops of radish from irradiated seeds were grown on an area of 1.5 ha at the Zarech'e state farm. (The yield obtained from an equal area sown with unirradiated seeds was used as control.) The harvest showed an 11% increase in the yield of tubers in comparison with the control, which amounted to about 1 ton of additional produce from one hectare. Besides, it was established that the presowing irradiation of radish seeds with 1 c causes early maturation by five to six days in comparison with the control.

In 1960, there was a large-scale commercial trial of the presowing irradiation of radish variety Rubin seeds (harvest of 1959, first class seeds with initial moisture content of 14%) by sowing it under hotbed conditions on the Vorontsovo state farm of Moscow province. One day before sowing, the seeds were irradiated by ^{60}Co gamma-rays at 1 c with the strength of 470 r/min, in the gamma-unit called GUBE (gamma unit for biological experiments). Of the total area of 160 hotbed frames with the crop, 80 were sown with irradiated and the other 80 with control seeds. After harvesting, the tubers were tied in bundles of ten each, to determine the average mass of one bundle. Consolidated data on the yield with dates of harvest and the average mass of the bundle are given in Table 4.2.

Table 4.2. Yield data of radish tubers of the variety Rubin grown from unirradiated and irradiated (1 c) air dried seeds in hotbeds

Date of harvesting	Yield of unirradiated radish		Yield of irradiated radish	
	No. of bundles	Average mass of one bundle, g	No. of bundles	Average mass of one bundle, g
Apr. 30	194	—	310	—
May 3	843	—	938	—
May 6	635	115	794	128
May 10	455	—	482	—
May 17	728	107	628	126

Unirradiated radish yielded 2,855 bundles and the average mass of each bundle was 111 g, while the irradiated radish produced 3,152 bundles with an average mass of 127 g. The total mass of unirradiated radish was 317 kg and that of the irradiated radish, 400 kg, which amounts to an increase of 26% in the yield. Maximum yield increases in terms of number of bundles, was obtained, as may be seen from the table, from the first harvest on April 30. This was due to the much earlier ripening of the irradiated radish tubers in comparison with the control.

Apart from yield estimate, the vitamin C content in the small and big tubers was also determined while experimenting with the irradiation of the radish variety Rubin grown in hotbeds. Analysis of the vitamin C content in radish tubers grown from the irradiated and unirradiated seeds showed that the presowing irradiation of seeds at 1 c increased the vitamin C content in tubers by 4–6 mg%.

	Large tubers	Small tubers
1 c	21.8	24.4
Control	17.8	18.4

Experiments on the presowing irradiation of air dried seeds of the radish variety Saks were conducted at the Institute of Biology of the LatvSSR [205]. Seeds were irradiated on the caesium unit with the gamma-equivalent of 0.5 g of radium and with the energy of 0.6 MeV. In the first experiment the seeds were sown seven months after the irradiation and immediately after irradiation in the second experiment. Stored seeds irradiated at 0.5 c showed an absence of stimulation and the tuber yield was 93% that of the control. Sowing seeds immediately after irradiation at 0.5, 0.3 and 0.1 c increased the tuber yield in comparison with the control by 19% at 0.5 c.

At L'vov University [78] experiments were conducted with sprouted radish seeds. For these the seeds were soaked in water for about two hours and stored in a moist condition at room temperature until the beginning of germination. After irradiation the sprouted seeds were mixed with dry sand and sown. The optimal dose of irradiation for the sprouted seeds was 0.5 c (Table 4.3). The tubers grown from irradiated seeds matured 10–12 days earlier than the control.

Table 4.3. Yield of radish tubers grown after the presowing irradiation of sprouted seeds

Radish variety	Irradiation dose, c	The number of hotbed frames in the experiment	The number of radish bundles obtained from one frame	Share in comparison with the control, %
Rozovyi with white tip	0.5	40	25.8 ± 0.2	115
	Control	40	22.4 ± 0.2	100
Rubin	0.5	100	24.8 ± 0.4	126
	Control	100	19.6 ± 0.7	100

Thus, the irradiation of sprouted seeds of the radish varieties Rozovyi with white tip and Rubin yielded 15 and 26% additional tubers in comparison with the control. A comparison of these data with data from the experiment with the radish variety Rubin grown in hotbeds (see Table 4.2) shows the closer values of the tuber yield increase. However, to get this effect, the irradiation dose of 0.5 c was required for the sprouted seeds and 1 c for dry seeds. (These experiments also prove the increased radiosensitivity of soaked seeds in comparison with the air dried seeds.)

The presowing irradiation of radish seeds of the variety Rozovyi with white tip at 1 c was also done at the Ural' branch of the USSR Academy of Sciences [208]. Strictly analogous growth conditions were created in all experiments. The first quality seeds were sown in 1 m^2 pots filled with absolutely homogenous soil. Each pot had ten rows of plants, of which five were sown with irradiated seeds alternating with five rows with control seeds. Light intensity uniformity was regulated with fluorescent and filament lamps. The plants in the pots with filament lamps were illuminated for 12 hours a day, but those with fluorescent lamps were lighted continuously.

In both illumination treatments the irradiation of radish seeds led to an increase in plant productivity: the tuber yield increased by 38.8% in plants grown under filament lamps and by 14.5% under fluorescent lamps in comparison with the control. This experiment is remarkable in the sense that it shows how significantly the stimulation effect can change depending on such diverse factors as the sources of illumination.

At the Ural' branch of the USSR Academy of Sciences, irradiating radish seeds of the variety Rozovyii with white tip increased the tuber yield by 17–28% in one experiment with much higher doses (10–25 c). The authors could not explain this considerable radiosensitivity increase in the radish [190].

T. Wash in Hungary, experimented with the presowing irradiation of radish and found that irradiation at 1.5 c increased the vitamin C content in tubers by 30%.

Indian researchers [277] studied the effect of x-radiation on radish seeds with a moisture content of 9.5% (the seeds were sown four hours after irradiation).

Stimulation was determined on the basis of the raw weight of the aerial parts and tubers. The data obtained by them are shown in Table 4.4.

The results of experiments on the presowing irradiation of radish seeds of different varieties are shown in Table 4.5. As is evident from the table, a number of researchers, working independently of each other, came to analogous conclusions about the stimulating effect of x-ray and gamma-irradiation at 0.5–1 c. With the irradiation of soaked seeds and the use of low doses, the optimal stimulation emerged at the very low dose of 0.5 c. Exceptions were the data from E.I. Preobrazhenskaya who got stimulation through irradiation at a considerably higher dose (10–25 c). The data on the disappearance of the stimulation effect during the long storage of irradiated seeds are of great interest.

Table 4.4. Average mass, length and volume of tubers

Irradiation dose, c	Average mass, g		Length, cm	Volume, cm
	aerial parts	tubers		
Control	71 ± 9	185 ± 9	16.0 ± 0.62	208 ± 13
0.1	68 ± 9	187 ± 18	16.9 ± 0.56	222 ± 20
0.5	76 ± 8	199 ± 16	15.7 ± 0.6	227 ± 18
1.0	79 ± 5	201 ± 13*	16.5 ± 1.17	245 ± 27*
0.1	70 ± 4	191 ± 14	16.4 ± 0.66	230 ± 18
0.5	73 ± 6	210 ± 20	17.2 ± 0.63	250 ± 23
1.0	77 ± 5	241 ± 15*	17.4 ± 0.40	276 ± 18*

*Positive differences were observed regarding the average mass and average volume of tubers irradiated at 1 c.

Table 4.5. Consolidated data on the presowing irradiation of radish seeds

Place of research	Year	Researcher	Dose of irradiation	Increase in yield against control, %	Change of biochemical composition	Enhancement of ripening, days
Institute of Biophysics of the USSR Academy of Sciences	1956	L.P. Breslavets, N.M. Berezina, G.P. Shchibrya and others	1 c	Increase in the length of roots under laboratory conditions, 19–33	—	—
Institute of Biophysics of the USSR Academy of Sciences, All-Union Research Institute of the Vitamin Industry	1962	N.M. Berezina, G.P. Shchibrya, M.L. Romanchikova	1 c	26	Increase in the vitamin C content by 4–6 mg%	5–6
Institute of Biology of the LatvSSR	1959	K.K. Roze, G.E. Kavatse	0.5 c	19	—	—
L'vov state university	1959	S.O. Grebinskii, V.G. Tsibukh	0.5 c (soaked seeds)	26	—	10–12
Ural' branch of the USSR Academy of Sciences	1963	V.N. Savin	1 c	11–38	—	—
	1963	E.I. Preobrazhenskaya	25 c	28	—	—
India	1970	—	5 Krad	13	—	—
Hungary	1971	T. Wash	1.5 Krad	—	Increase in the vitamin C content by 30%	—

Carrot. Experiments on carrot were conducted by researchers of the Institute of Biophysics of the USSR Academy of Sciences in collaboration with researchers of the All-Union Research Institute of the Vitamin Industry. Air dried seeds of the carrot varieties Nantskaya, Shantane, Nesravnennaya. Losinoostrovskaya and Moskovskaya zimnyaya were subjected to different presowing treatments to increase the germination energy, germination percentage and yield. These treatments included soaking in malt extract, in solutions of potassium bromide, ammonium molybdate, manganese sulfate, nicotinic acid, thyamine and also x-ray irradiation. The best results with respect to yield and carotine content in the tubers were obtained from x-irradiated seeds and those soaked in the potassium bromide solution. The yield increases from these treatments were 35% and 42% respectively. Of these two seed treatments, preference should be given to x-irradiation because soaking of seeds in potassium bromide solution is considerably more troublesome than irradiating air dried seeds.

Taking this into consideration, we began to experiment on the presowing irradiation of carrot. Air dried seeds were x-irradiated at 0.25, 0.5, 1, 2, 4 and 8 c at the following operating conditions: tube voltage 170 kV, current strength 6 mA, distance from the source—60 cm, strength of the dose—52.1 r/min using a cardboard filter. Roots developed from seeds germinated in petri dishes were measured on the second, fourth and sixth day (Table 4.6).

As evident from the table, the number of sprouted seeds increases with irradiation in doses of 1–8 c; root growth is stimulated with irradiation at 0.25–8 c. The carrot seeds were found to be very radiosensitive to the effect of x-irradiation. The stimulation effect appeared in a wide range of doses. After obtaining the stimulation effects of irradiation with doses of 2, 4 and 8 c under laboratory conditions, we planned experiments under field conditions at the Vorontsovo state farm of Moscow province, in which we tried

Table 4.6. The number of sprouted seeds and the length of the roots of carrot sprouts after the presowing x-ray irradiation of seeds

Dose of irradiation, c	No. of sprouted seeds in comparison with the control, %	Length of roots, mm	Length of roots in comparison with the control, %
Control	100	34.9	100
0.25	100	38.9	111
0.5	100	39.8	114
1.0	147	42.1	120
2.0	136	43.3	124
4.0	160	40.0	116
8.0	113	39.1	112

the irradiation effects at doses of 1–8 c. The results of the experiments conducted in six replications on plots with an area of 100 m^2 are given in Table 4.7.

As evident from the given data, the presowing irradiation of air dried carrot seeds at 4 c made it possible to increase the tuber yield by 26% in comparison with the control. Carotine content was determined in tubers grown from irradiated seeds (Table 4.8).

Since such a significant carotine reserve due to the effects of irradiation can be of great value to the vitamin industry, we conducted large-scale commercial sowings of irradiated carrot seeds of the variety Nantskaya on the production farm of the Krasnodar vitamin plant [96]. Irradiated seeds were sown in 1957 over an area of 5.5 ha and on 29.5 ha in 1959. The tuber yield for two years increased by 29 and 30%, while the carotine content exceeded the control by 3 and 12%. At the request of the Krasnodar vitamin plant in 1964, we irradiated 720 g of carrot seeds of the variety Nantskaya to be sown on the entire production area of the plant.

In 1961, irradiated carrot seeds of the variety Nesravnennaya were sown over an area of eight hectares on the Zaokskii state farm of Moscow province. The seeds were irradiated with ^{137}Cs gamma-rays on the mobile unit

Table 4.7. Yield of carrot tubers of the variety Nantskaya grown from the irradiated seeds

Dose of irradiation, c	Yield, q/ha	Percentage in comparison with the control
Control	182.0	100.0
1	209.3	114.4
2	225.0	123.0
4	229.5	126.0
8	223.5	122.2

Table 4.8. Total content and reserve of carotine in the carrot tubers grown from irradiated seeds

Dose of irradiation, c	Carotine content		Carotine reserve	
	mg%	%	kg/ha	%
Control	18.50	100.0	3.02	100.0
1	19.38	104.5	4.06	134.5
2	21.28	115.5	4.76	157.7
4	20.23	109.4	4.70	156.0
8	18.06	97.6	4.14	137.0

GUPOS at 2.5 and 4 c with the strength of 720 r/min. Results of the experiment showed a decrease in the germination of seeds irradiated at 4 c in comparison with the control and the thinning of plants to a certain extent: plant density in this treatment was 82.3% of the control; the data on the tuber yield are given in Table 4.9.

Table 4.9. Tuber yield of the carrot variety Nesravnennaya grown from the seeds irradiated with gamma-rays

Dose of irradiation, c	Yield, q/ha	Percentage against the control
Control	330	100.0
2.5	448	135.4
4.0	400	121.7

While harvesting the yield on the collective farm Zaokskii, the quantity of standard and sub-standard tubers grown from the irradiated and unirradiated seeds was also taken into consideration. These studies showed that one reason for the increase in the yield of the Nesravnennaya carrot variety was an increase in the number of large standard tubers as a result of seed irradiation at 2.5 c. Preharvest biochemical analysis, in comparison with the control, showed that the vitamin C content in the tubers grown from seeds irradiated at 2.5 and 4 c was higher by 8 and 11.9 mg% respectively. Carotine also increased by 3.49 and 4.3 mg% respectively. It must be mentioned that, in contrast with the maximum yield increase, the optimal biochemical changes in the plants, particularly the increased ascorbic acid and carotine, were also observed on irradiated seeds in much higher doses.

Comparing the effects of the presowing irradiation, as observed in different varieties of carrot, we see that the best results were obtained in the variety Nantskaya. Along with the increased tuber yield, we observed a greater increase in the carotine content in them than in carrot tubers of the remaining varieties under study, in which the increase in the reserve carotine was essentially due to the yield increase. N.G. Zhezhel and R.K. Baranov obtained a 35% increase in the carrot tuber yield with a 11–51% increase of carotine content.

The presowing irradiation of carrot seeds was also done in the Department of Plant Physiology of L'vov State University [78]. In contrast with our experiments, instead of air dried seeds they irradiated sprouted seeds of the carrot variety Shantane. The irradiation of sprouted seeds at 0.5 c increased the yield by 48% in comparison with the control. The results obtained, as well as the data regarding the irradiation of sprouted radish seeds, affirm the considerably greater radiosensitivity of sprouted seeds than the air dried seeds. It is also shown that greater stimulation effectiveness is obtained by irradiating sprouted seeds rather than dry ones.

The effects of irradiating carrot seeds of the variety Nantskaya 14 were tested for three years by the state commission on varietal trials of agricultural crops (Gossortoset'). On this basis presowing irradiation was recommended for Leningrad province (as per data from the Agrophysics Institute).

Table 4.10 shows the results of experiments on the presowing irradiation of the carrot seeds.

Cabbage. The first experiments on the irradiation of cabbage seeds were conducted in 1914 by Köernike who established that low doses of x-radiation accelerated the growth and development of cabbage. Laboratory research on the presowing irradiation of seeds of white head cabbage variety Kolkhoznitsa, conducted at the Institute of Biophysics of the USSR Academy of Sciences, has shown the stimulation of the x-ray radiation of seeds at 1–2 c (voltage 188 kV, strength of the current—5 mA, cardboard filter, distance from the tube—60 cm and strength of the dose—42.5 r/min.). Field experiments to verify the laboratory data were conducted by L.P. Breslavets at the Gribovskaya vegetable breeding station. Her results showed that the presowing irradiation of air dried cabbage seeds increases the yield of heads by 19% in comparison with the control.

In 1961, cabbage seeds of the variety Slava were irradiated on Zaokskii state farm of Moscow province. The seeds irradiated by ^{60}Co and ^{137}Cs gamma-rays at 2 c, were sown in an open field on an area of 8 ha. When assessing the germination percentage, no differences were noticed between irradiated and unirradiated seeds. However, subsequent observations showed that the growth and development of the cabbage seeds in irradiated plots was more intensive than that in the control. In seedbeds sown with irradiated seeds, the number of economically suitable seedlings by the time of transplantation was more (19–17 seedlings per 1 m^2) than in the control (11 seedlings) [38].

On August 17 the cabbage heads (200 heads in every experimental treatment) were assessed for their development by dividing them into three categories: fully developed, underdeveloped and heads in embryonic condition (Fig. 4.2). The percentage distribution of plants on the basis of the head development was as follows:

	Fully developed heads	Underdeveloped heads	Heads in embryonic condition
Control gamma-irradiation	17	26.5	56.5
^{60}Co	49.5	38	12.5
^{137}Cs	58.5	34	7.5

In this way the data obtained show that the presowing irradiation of cabbage seeds of the variety Slava accelerates the plant development and the formation of cabbage heads.

Table 4.10. Consolidated data on the presowing irradiation of carrot seeds

Place of research	Year	Researcher	Dose of irradiation, c	Increase of yield in comparison with the control,%	Change of biochemical composition
All-Union Research Institute of Vitamin Industry	1956	G.I. Shchibrya	2.5 + potassium bromide	35	Increase of carotine content
Institute of Biophysics of the USSR Academy of Sciences and All-Union Research Institute of Vitamin Industry	1956	L.P. Breslavets, N.M. Berezina, G.I. Shchibrya and others	4	26	Increase in the vitamin C content by 8–14 mg%, carotine by 3–4 mg%
Institute of Biophysics of the USSR Academy of Sciences and Krasnodar vitamin plant	1958	N.M. Berezina, G.I. Shchibrya, A.I. Solodukhin, V.B. Kiryukhin	2.5	34	Increase in carotine content by 3–12%
L'vov State University	1959	S.O. Grebinskii, V.G. Tsibukh	2.5, 0.5 (sprouted seeds)	48	—
Krasnodar vitamin plant	1964	T.P. Zhdanova, A.I. Solodukhin, V.N. Pavlov	2.5	7	Increase in carotine content by 22%
Institute of Biophysics of the USSR Academy of Sciences; Zaokskii state farm of Moscow province	1963	N.M. Berezina, R.R. Riza-Zade	2.5	—	Increase in carotine content

Leningrad Agricultural Institute	1970	N.G. Zhezhel', R.K. Baranov, B.V. Kozyreva. V.P. Pavlov	—	—	Increase in carotine content by 11–56%
Institute of Breeding and Genetics of the Academy of Sciences of AzSSR	1968	Zh.B. Brazhnikova	Variety Nantskaya 2.5–5	22–25	Increase in carotine content by 18–22%
			Variety Apsheronskaya 2.5–5	24–28	Carotine content higher by 1–14%
L'vov State University	1968	S.O. Grebinskii	—	20–30	—

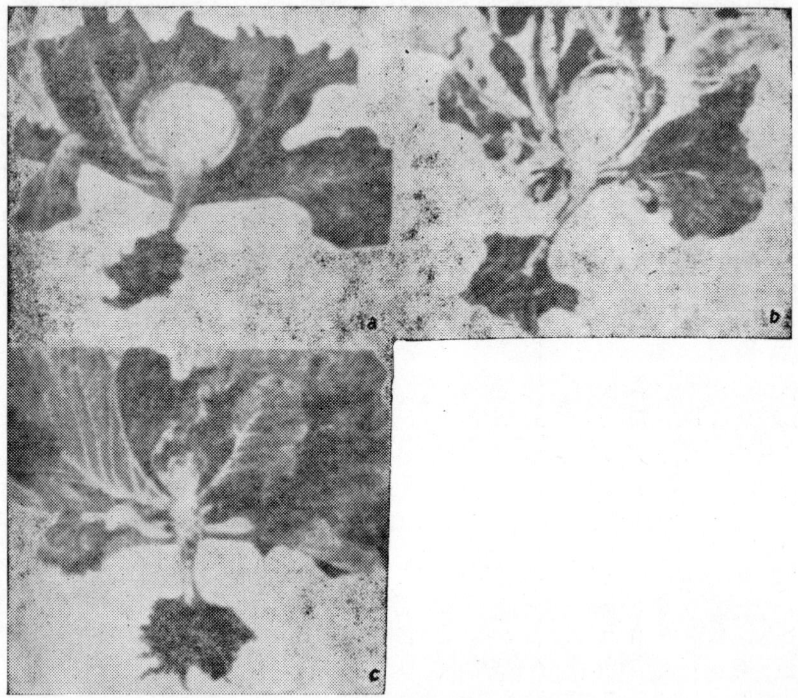

Fig. 4.2. Development of cabbage head with the presowing irradiation of seeds: *a*—fully developed heads; *b*—underdeveloped heads; *c*—heads in the embryonic condition.

During the yield assessment of heads on September 29–30, the increase which was caused by the irradiation of seeds with ^{60}Co and ^{137}Cs gamma-rays was observed to the tune of 20 and 21% respectively.

The vitamin C content of the seedlings and heads was estimated during the vegetative period (Table 4.11).

It is evident from the table that in the initial development stages the vitamin C content is very high in the seedlings; the vitamin C content in seedlings grown from the ^{137}Cs and ^{60}Co gamma-irradiated seeds was 46 and 69 mg% higher than in the control. With the growth and development of seedlings their vitamin C content is reduced in an orderly way, however, in plants grown from irradiated seeds, it remains at a much higher level at all times than in the control. The analogical regularity or order was also noticed during the development of the cabbage heads. The ascorbic acid content decreases orderly with the ripening of the heads, however, at maturity all plants grown from the ^{60}Co and ^{137}Cs gamma-irradiated seeds contain 9.4 and 12.2 mg% of vitamin C more than the control plants.

The data about the presowing irradiation of seeds of white headed (table purpose) and fodder cabbage obtained at the Ural branch of the USSR Academy of Sciences are presented in Table 4.12.

A significant increase in the white headed cabbage yield was obtained with doses of 0.25, 0.5 and 1 c and in the fodder cabbage yield with doses of 0.5, 2 and 4 c, which shows the much higher radiosensitivity of the fodder cabbage.

Table 4.11. Vitamin C content in cabbage after the presowing irradiation of seeds at 2 c, mg%

Treatment	Seedlings		Heads			
	June 9	June 29	August 17	September 9	September 22	September 30
Control	400.2	193.2	87.8	81.6	73.1	71.0
Gamma-rays: ^{60}Co	469.2	285.0	99.7	89.6	85.5	80.4
^{137}Cs	446.2	248.0	102.4	83.5	84.5	83.2

Table 4.12. Effects of the presowing irradiation of seeds on the yields of white headed (table purpose) and fodder cabbage

Dose of irradiation, c	White headed or table purpose cabbage		Fodder cabbage	
	Average mass of one head, g	Percentage against the control	Average mass of one head, g	Percentage against the control
Control	520	100.0	1,770	100.0
0.25	630	121.0	1,830	103.2
0.5	600	115.2	1,990	112.5
1.0	580	111.5	1,920	108.5
2.0	610	117.2	1,960	110.6
4.0	560	107.8	1,980	111.9
8.0	590	113.6	1,830	103.2

G.I. Satalkina [212] conducted presowing irradiation of the seeds of cabbage varieties Ditmorskaya rannyaya and Nomer pervyi Gribovskii. The data obtained by her are interesting from the viewpoint that in Krasnodar territory conditions, the stimulation doses for cabbage were considerably much higher than in the central non-chernozem belt, particularly in the dose range of 3–8 c. Table 4.13 shows data about the effect of a wide range of irradiation doses on the yields of cabbage varieties Nomer pervyi Gribovskii and Ditmorskaya rannyaya during 1965–1966.

Table 4.13. Effects of the presowing gamma-irradiation of seeds on cabbage yield

Dose of irradiation, c	Nomer pervyi Gribovskii				Ditmorskaya rannyaya			
	Total yield		Yield of early harvests		Total yield		Yield of early harvests	
	q/ha	%	q/ha	%	q/ha	%	q/ha	%
Control	328	100.0	61.1	100.0	299	100.0	152.5	100.0
1	336	102.4	67.8	110.9	313	104.0	193.4	126.8
3	355	108.2	83.8	137.2	337	112.7	236.7	155.2
4	370	112.8	91.5	149.8	340	113.7	213.4	139.9
8	372	113.4	76.4	125.0	347	116.0	176.0	115.4
16	335	102.3	56.9	93.1	294	98.4	128.5	84.3

These data have theoretical interest because the range of stimulation doses for the two irradiated cabbage varieties was very wide, i.e., from 1 to 8 c and, in certain cases, even up to 16 c. This confirms the observations of E.I. Preobrazhenskaya and N.M. Berezina that highly radiostable plants, which also include cabbage as a representative of the family Cruciferae, have a much wider range of stimulation doses than radiosensitive plants. The practical importance of the presowing irradiation of cabbage increases because, under the action of the stimulation irradiation dose, the vitamin C content increases significantly in the cabbage heads. In the leaf formation phase of the control plants (in seedlings) the vitamin C content was 104.7 mg%. For seeds of the cabbage variety Slava irradiated at 3 c, the vitamin C content was 121.5 mg%. The irradiation of cabbage seeds in the harvest maturity phase increased the vitamin C content by 6–8 mg%, not only in leaves but also in the heads.

Results of experiments on the presowing irradiation of cabbage are presented in Table 4.14.

Cucumber. In the Institute of Biophysics of the USSR Academy of Sciences, air dried cucumber seeds of the varieties Klinskie, Nezhinskie and Vyaznikovskie were irradiated. The effects of a wide range of doses, i.e., 0.1–8 c was tested under laboratory conditions. With the irradiation of seeds at 0.1 c, the growth and development of plants was stimulated in the initial stages of their ontogenesis. However, by the time of flowering these plants were no different from the control plants. Seeds irradiated with 0.3 c exhibited a stable stimulation which was maintained throughout the entire vegetative period and was manifest in the much faster development of leaf blades and *whisk* formation as well as much earlier flowering and fruit set. The irradiation dose of 4 c stunted the initial stages of plant development, but in their later growth, these plants recovered and made good the initial loss and by the time of flowering they surpassed the control plants.

Our laboratory data were verified by several researchers in field experiments at the Gribovskaya vegetable station, where seeds of the cucumber variety Vyaznikovskie irradiated with 0.3 c were sown here. The green mass yield increased by 30% with much earlier fruit maturity.

During the last few years, a number of positive data about the presowing irradiation of cucumber were obtained at the Gorky university named after N.I. Lobachevskii. It was found that the stimulation dose of gamma-radiation is not uniform under different conditions for growing cucumber (in the open field and in glasshouse conditions). In experiments with cucumbers of the glasshouse variety Murmanskii mnogoploidnyi (in the state farms Gor'kovskii, Zhdanovskii and Vpered of Gorky province) the dose of 1 c has shown a stimulation effect. In experiments with cucumber in the open fields and in hotbeds, the stimulation effect was shown with much lower doses of 0.2–0.3 c. The increased yield of green mass (tender,

Table 4.14. Consolidated data on the presowing irradiation of seeds of white headed and fodder cabbage

Place of research	Year	Researcher	Dose of irradiation, c	Increase of yield in comparison with the control, %	Change of biochemical composition	Enhancement of development
Institute of Biophysics of the USSR Academy of Sciences and Gribovskaya experimental station	1953	L.P. Breslavets	2	19 (white headed cabbage)	—	Enhancement of ripening
Institute of Biophysics of the USSR Academy of Sciences and Zaokskii state farm of Moscow province	1961	L.P. Breslavets, R.R. Riza-Zade, E.I. Korneva	Gamma-rays ^{60}Co ^{137}Cs	20 (white headed cabbage) 21 (white headed cabbage)	Increase of vitamin C content in seedlings by 46–49 and in the heads by 9.4–11.2 mg%	Enhancement of ripening
Ural branch of the USSR Academy of Sciences	1959	N.A. Poryadkova, N.M. Makarov, N.V. Kulikov	2	17 (white headed cabbage) 10 (fodder cabbage)	—	—
Krasnodar Agricultural Institute	1970	G.I. Stalkina	3–8	Nomer pervyi Total yield—13 Early harvest—49 Ditmorskaya Total yield—16 Early harvest—55	Increase of vitamin C content in the leaves by 16 and in the heads by 6–8 mg%	Enhancement of ripening

immature fruits) at the state farm Vpered of Gorky province over three years is shown in Table 4.15.

The presowing irradiation of the seeds of cucumber varieties Nezhinskie and Kirovobadskie was done at the Institute of Genetics and Physiology of the Academy of Sciences of the AzSSR [239]. Wide range of doses for irradiating cucumber (0.5–40 c) was tried in field experiments during 1958. Phenological observations showed that under Azerbaijan conditions a dose of 40 c considerably retarded the growth and development of plants. Optimal stimulation effects on the development and growth were produced by irradiating with 4 c: the vegetative period of the cucumber variety Nezhinskie was shortened by 11 days and that of the variety Kirovobadskie by 7 days. This irradiation dose made it possible to get a maximum increase of 18% in the yield of green mass (tender fruits) under field conditions (Table 4.16).

We must pay attention to the observations which showed that a high dose of irradiation (40 c) prolonged the period of vegetative growth and caused the formation of new stolons and fruit set in many small fruits right up to the onset of frost. The most noticeable enhancement in plant development and passage of individual developmental phases takes place with irradiation at 4 c at the initial stages, i.e., from seedling emergence to flowering.

Table 4.15. Results of the presowing irradiation of cucumber seeds in Gorky province

Year	Dose of irradiation, c	Yield of green mass (tender fruits), %
1963	Control	100.0
1963	0.3	148.0
1963	0.2	119.0
1964	0.3	148.1
1965	0.3	123.6
1965	0.3	121.0
1965	0.3	123.7
1965	0.8	133.0

Table 4.16. Cucumber yield of the variety Nezhinskie grown from irradiated seeds

Dose of irradiation, c	Yield, q/ha	In percentage of the control
Control	275.0	100.0
4	325.4	118.0
16	299.2	108.0
20	286.4	104.0
40	175.7	64.0

Among the morphological changes in cucumber due to the effect of irradiation, we must take note of the capability of plants to form a large number of lateral stolons (7–9 per plant) in comparison with the control (3–4).

The stimulation effect of irradiation at 4 c was tested on commercial cucumber crops in the V.I. Lenin collective farm of the Astaurinskii region. In the crop of the cucumber variety Nezhinskie the yield of green, tender fruits increased by 28 q/ha over the control, i.e., the increase was only 13%. Biochemical analysis showed that with an increased irradiation dose there is an increase in the vitamin C content in the cucumbers.

Sprouted seeds of the cucumber variety Semennikovskie were irradiated at L'vov University. The stimulation effects of irradiation emerged with dose of 0.25–0.5 c. The yield of cucumbers grown in hotbeds from seeds irradiated at 0.25 and 0.5 c increased by 19 and 38% respectively. The results so obtained show the presence of a very wide range of stimulation doses for cucumber, which changes depending on the condition of the seeds and the growth.

A.G. Sidorskii noticed that when cucumber seeds are irradiated with a dose to stimulate plant growth and development the quantity of female flowers increases in the second half of the vegetative growth period. This interesting phenomenon can be considered as one factor responsible for the increased yield of green produce or tender fruits.

In the Institute of Agriculture, Kirgiz SSR [222] field experiments were conducted on the presowing irradiation of the cucumber variety Nemchinovskie with the doses of 1, 5, 10 and 15 c. As a result of irradiation at 10 and 15 c, the yield of green produce (tender fruits) increased by 9.7 and 14.9% respectively. Such high stimulation doses deserve further careful study. One explanation for the high level of stimulation doses under the conditions at Kirgiz SSR may be that natural radioactivity level of soils in the Chuiskaya valley, where the experiment was conducted, is very high, this may induce an adaptation to the effect of gamma-rays for plants grown from locally reproduced seeds.

Under open field conditions in the MSSR, the presowing irradiation of the cucumber varieties Teplichnyi 40, Druzhnyi 85, Syurpriz 66, Gibrid 65 and Odnostebel'nyi 33 was done on a LMB-4-1M gamma-unit with doses of 0.3, 0.5, 0.75 and 1 c.

As a result of the irradiation of seeds at 0.5 c, the hybrids Teplichnyi 40 and Druzhnyi 85 produced up to a 5% increase in yield. The optimal dose for the hybrid Syurpriz 66 was 0.75 c, at which there was an 11% increase in the yield of green produce (tender fruits) with much earlier ripening. The output of marketable produce from irradiated plants was 6–8% higher than in the control. The yield of cucumbers of the variety Odnostebl'nyi 33, after seed irradiation at 0.5 c, exceeded the control yield by 10–12%, which comprised 1.6–2 kg of additional produce in terms of output from 1 m^2.

Foreign data presented by the Canadian scientist MacKey at the Geneva conference show that the presowing irradiation of cucumber seeds in Canada with doses of 0.3–1 c makes it possible to obtain a 40–90% increase in the yield of green produce (tender fruits) with much earlier ripening in comparison with the control.

Generalizing the results of the presowing irradiation of cucumber seeds, we see that the value of the optimal stimulation dose for this crop changes sharply. Depending on the growth conditions and the different geographical zones, it varies within a range of 0.2–15 c (Table 4.17).

Potato. As shown by much research, potato belongs to the category of highly radiosensitive crops. The stimulation dose for potato tubers lies within a range of 0.1–0.5 c.

At the Institute of Potato Cultivation of Moscow province, experiments on the presowing irradiation of tubers and seeds [79] were conducted on potato varieties Lorkh (medium maturity) and Priekul'skii (early). The tubers were irradiated with doses of 0.1, 0.15, 0.5, 1, 3, 5 and 8 c on the unit GUPOS with the strength of dose at 60 r/min. Five days after irradiation the tubers were planted in the soil (peat-bog and sodpodzolic soils). The number of sprouted eyes (beds) was counted before planting.

The experiments showed that the tubers did not sprout after irradiation at 5–8 c. Sprouting was considerably delayed and the number of sprouted buds decreased in tubers which were irradiated at 1–3 c. Irradiation at 0.15–0.5 c reduced the period from planting to sprout emergence by three to four days and increased the number of sprouted buds in one tuber. For tubers irradiated at 0.15, 0.3 and 0.5 c, there was increase in plant height, number of stems, root mass, area of assimilation surface of the leaves and tuber yield.

Presuming that one of the factors that stimulated the growth and development of plants, grown from the irradiated tubers can be an increase in the mitotic index, the authors considered the intensity of cell division in the sprouts of 10 mm irradiated tubers. At 0.3 c the intensity of cell division increased in the shoots of the variety Lorkh by 53.3% and by 54.5% in rootlets when compared with the control. An analogous change in mitotic activity was also observed in the variety Priekul'skii. A dose of 1 c weakened mitotic activity.

The determination of nucleic acid content in the sprouts of irradiated tubers showed that due to the presowing irradiation of tubers with 0.3 c the synthesis of nucleic acids increased in the plants, but a dose of 1 c which depresses growth and development, reduced synthesis. The dose of 0.3 c resulted in an increase in RNA synthesis by 19% and that of DNA by 83% in comparison with the control, which is evidently linked with the intensified division of cell nuclei in the plants under the effects of irradiation. As a result of the irradiation of potato tubers with 0.3 c the content of auxins in the sprouts increased more than tenfold.

Table 4.17. Consolidated data on the presowing irradiation of cucumber seeds

Place of research	Year	Research	Dose of irradiation	Increase in yield in comparison with the control, %	Change of biochemical composition	Enhancement of maturity, days
Institute of Biophysics of the USSR Academy of Sciences	1955	L.P. Breslavets, N.M. Berezina	0.3–0.4 c	23 18	—	6
Gribovskaya vegetable breeding station	1956	—	0.3 c	30	—	—
Institute of Breeding and Genetics of the Academy of Sciences of AzSSR	1959	A.I. Khudadatov	4 c	18	Increase in the vitamin C content up to 21 mg %	9–11
L'vov State University	1959	S.O. Grebinskii, V.G. Tsibukh	0.25–0.5 c	18 19	—	—
V.I. Lenin collective farm of Astaurinskii region of AzSSR	1960 1961	A.I. Khudadatov	4 c	13	—	7–10
Gorky State University	1966	V.A. Guseva, A.G. Sidorskii	0.2–0.8 c	19–48	—	Enhancement of maturity
Canada	1969–1970	MacKey	0.3 c	40–90	—	-do-
Kirghiz SSR	1972	A.S. Sultanbaev, L.A. Sergeeva	10–15 c	9.7–14	—	—

The effects of irradiation on the intensity of photosynthesis and outflow of plastic substances from the leaves of irradiated potato were studied on the same experimental material using the ^{14}C isotope. Photosynthesis in plants grown from tubers irradiated at 0.3 c went on at a much higher level than in the control. The photosynthesis intensity with this dose of irradiation increased, depending on the time of the day, by 18.2–40.2% in comparison with the control. After 24, 48 and 72 hours post-exposition, cuttings (chips) were taken from the labeled leaf to determine the outflow of plastic substances. Investigation results showed that the outflow of plastic substances in plants grown from tubers irradiated with 0.3 c two days after labeling the leaf was almost two times more intensive than in the control plants. Irradiation caused a more intensive accumulation of assimilates in the potato tubers. For example, 48 hours after labeling the leaf, their content was 77.5% higher than in the control.

To reveal the intensity of mineral nutrition of plants grown from irradiated tubers, the assimilation of phosphorus was observed using the isotope method. Research showed that phosphorus intake into plants grown from tubers irradiated at 0.1–0.3 c takes place more intensively than in the control (Table 4.18).

Table 4.18. Effects of the preplanting gamma-irradiation on the uptake of radioactive phosphorus by potato plants of the variety Lorkh

Dose of irradiation, c	Content of ^{32}P per gram of dry matter					
	Budding		Flowering		10 days after flowering	
	Thousand imp./min	%	Thousand imp./min	%	Thousand imp./min	%
Control	64.3	100	55.3	100	65.1	100
0.1	77.7	121	57.5	122	77.8	120
0.3	92.9	145	88.5	160	94.6	145
1.0	45.5	70	46.3	84	54.2	835

The amount of starch and vitamin C was determined in average samples during harvesting (Table 4.19).

As evident from the table, the doses of gamma-irradiation of 0.15, 0.3, and 0.5 c improve the tuber quality.

Planting irradiated tubers of potato in different soils, V.S. Serebrenikov noticed that a maximum increase of potato yield was obtained from the variety Priekul'skii grown in peat and sandy loam soils. When planting tubers of this variety in sandy loam soil, the optimal dose of irradiation was

Table 4.19. Effects of the preplanting gamma-irradiation of potato tubers on the quality of tubers (peat-bog soil)

Dose of irradiation, C	Lorkh		Priekul'skii	
	Starch, %	Vitamin C, mg%	Starch, %	Vitamin C, mg%
Control	13.2	13.3	12.3	13.4
0.1	13.6	15.0	—	—
0.15	14.5	18.5	12.8	14.6
0.3	15.2	19.4	13.3	17.7
0.5	14.4	16.5	14.1	19.1
1.0	11.1	11.8	10.1	13.9

0.3 c, but 0.5 c in peat soils. The mathematical analysis of data on the increased potato yield showed they were totally significant and reliable. Results of the yield estimate are shown in Table 4.20.

Generalizing the data obtained as per different indices, we see that the factors that contribute toward increased potato yields are the increased intensity of photosynthesis and mitotic activity, much faster outflow of plastic substances and the improved nutrition of sprouts because of greater uptake of mineral salts by the plants grown from tubers irradiated with 0.1–0.3 c.

Moreover, the preplanting irradiation of tubers was tried for two years under large-scale production conditions in the Orlovskoe experimental-cum-demonstration farm and in the Kolomenskoe and Put' Novoi Zhizni state farms (Table 4.21).

To find the optimal storage periods of irradiated seeds, they were irradiated at 0.1–0.15, 0.3, 0.5 and 1 c on January 8, March 10, April 13 and May 19, 1960. Irradiation on January 8 and March 10 with 0.3 and 0.5 c adversely affected the productivity of plants. In this case there was vigorous and premature (before time) sprouting of the buds and, as a result, by the time of planting the tubers had already lost their turgor and nutrient reserves to a considerable extent, and the sprouts broke down.

In the case of tubers irradiated 36 days before planting (April 13) and immediately before sowing (May 19), the optimal yield increase was found with the dose of 0.3 c, but it was higher for tubers irradiated before planting (Table 4.22).

At the Institute of Potato Cultivation, along with the tuber seeds of the potato variety Smyslovskii and hybrids 2x and 12x from self-pollination were also irradiated. The seeds were irradiated by gamma-rays with doses of 0.5, 0.1, 10, 15, 30, 45 and 60 c with the strength of the dose at 340 r/min.

Table 4.20. Yield of different varieties of potato grown from tubers irradiated with gamma-rays

	Lorkh		Priekul'skii					Lorkh	
	Sandy loam soil		Peat		Sandy loam soil			Irrigated plot (sandy loam soil)	
Dose of irradiation, c	Yield, q/ha	In percentage of the control	Yield, q/ha	In percentage of the control	Yield, q/ha	In percentage of the control	Error, % *	Yield, q/ha	In percentage of the control
Control	164.0	100	157.6	100	133.3	100	2.6	280.5	100
0.1	190.7	117	—	—	154.4	116	3.2	—	—
0.15	193.3	118	175.6	112	165.3	124	1.7	364.3	126.8
0.3	194.6	119	192.0	122	180.6	136	0.8	328.6	117.0
0.5	177.1	108	226.0	144	159.8	120	1.6	314.0	112.5
1.0	156.5	96	160.6	102	138.5	104	2.6	—	—

*Five replications; plot area 50 m^2.

Table 4.21. Yield and quality of potato grown under production conditions from tubers irradiated by ^{60}Co gamma-rays

Dose of irradiation, c	Yield, q/ha	In percentage of the control	Starch content, %
	Experimental farm Orlovskoe, 1962		
Control	159.2	100	16.9
0.3	181.5	114.0	18.4
	Collective farm Put' Novoi Zhizni, 1963		
Control	106.2	100	17.5
0.3	123.3	115.8	18.7
	State farm Kolomenskoe, 1962		
Control	112.3	100.0	17.1
0.3	134.5	119.7	18.2
	State farm Kolomenskoe, 1963		
Control	113.8	100.0	18.0
0.3	136.9	121.9	19.1

Most of the seeds irradiated at 60 c did not germinate and the plants which did emerge also perished completely. Of those irradiated with 45 c, 30% of the emerged plants died. Doses of 30, 15 and 10 c highly depressed the growth of the seedlings. The irradiation of dry seed at 1 c and that of the sprouted ones at 0.5 c caused the much earlier emergence of seedlings which also developed much faster. The potato yield grown from seeds irradiated in these doses surpassed the control by 35–50%. These data enable us to conclude that potato seeds are more radiosensitive than tubers: the stimulation dose for seeds is 1 c, but 0.3 c for the tubers.

Potato tubers were also irradiated at the Institute of Biology of Academy of Sciences of Latvian SSR [205]. The data obtained there showed that the irradiation of potato tubers with doses of 0.02–0.4 c, stimulated the growth and development of the potato and increased its yield by 8–10%. The increased yield potential was a result of the increased number of tubers per plant.

Experiments on the preplanting irradiation of potato were also conducted at the Institute of Biophysics of the USSR Academy of Sciences. Tubers of the potato variety Berlikhengen stored in an underground godown at 3–5°C were irradiated on the unit GUBE by ^{60}Co gamma-rays with doses of 0.1, 0.25, 0.5, and 2 c with the strength of 60 r/min. Five days after irradiation the tubers were planted. Between irradiation and planting the potatoes were stored at 18–20°C. The effects of irradiation were assessed in the field experiment at the Vorontsovo State farm, the experimental base of the

Table 4.22. Yield of potato depending on the time of gamma-irradiation of tubers (peat-bog soil, variety Lorkh, data of 1960)

Dose of irradiation, c	January 8		March 10		April 13		May 19	
	Yield, q/ha	In percentage of the control	Yield, q/ha	In percentage of the control	Yield, q/ha	In percentage of the control	Yield, q/ha	In percentage of the control
Control	130.0	100	130.0	100	130.0	100	130.0	100
0.1	139.6	107.5	140.7	108.3	147.3	113.1	153.4	118
0.15	132.3	102.2	147.5	113.2	156.4	121.0	170.0	131
0.3	126.4	97.3	141.4	108.8	168.4	129.3	186.0	143
0.5	128.4	98.7	146.2	112.5	152.1	117.0	151.6	117
1.0	128.9	99.5	127.2	97.6	122.3	94.0	120.6	93

Note: Basal dressing with: compost 20 ton/ha, ammonium nitrate 1.5 q/ha, superphosphate 2.0 q/ha, potassium chloride 1.5 q/ha.

Institute of Vitamin Industry in the Moscow province. The area of the plots was 90 m² and the experiment was conducted in three replications. The dose of 2 c caused a certain reduction in the germination. Slight flowering enhancement was observed with the irradiation in the dose of 0.1–0.5 c, 2 c delayed flowering (Table 4.23).

Data on the yield of potatoes grown from irradiated tubers are shown in Table 4.24.

The stimulation effect of such a high dose as 2 c on the development of the potato variety Berlikhengen, as was revealed, made it necessary to conduct additional investigations. The experiment on potato irradiation was repeated. The tubers were irradiated with doses of 0.5 and 2 c in November and were planted in a luminostat simultaneously with the control. After one month considerable delay was noticed in the sprouting of tubers that were irradiated with 2 c in comparison with the control and the tubers irradiated with 0.5 c. Analyzing the reasons for the growth retardation it was found that the bud development in tubers irradiated with 2 c began simultaneously with the bud development in the control tubers. Then, the sprouted buds were isolated from the mother tuber by a cork layer and developed a strong

Table 4.23. Assessment of the initiation of flowering in potato depending on the gamma-irradiation dose of tubers

Dose of irradiation, c	Number of plants assessed	Flowering plants	
		Number	%
Control	350	7	2.0
0.1	308	10	3.2
0.25	287	8	2.8
0.5	313	6	1.9
2.0	275	1	0.36

Table 4.24. Yield from potatoes grown from tubers irradiated by ^{60}Co gamma-rays (average of three replications)

Dose of irradiation, c	Yield, kg	In percentage of the control	Significance of difference
Control	63.4 ± 3.1	100	—
0.1	74.7 ± 10.4	117.3	1.05
0.25	86.9 ± 9.7	137.1	2.3
0.5	88.4 ± 3.6	139.4	5.3
2.0	98.5 ± 3.6	155.4	7.5

root system separately and considerably earlier than the control; in the period of this transformation the buds of tubers irradiated with 2 c were more retarded in growth and development than the buds of control tubers developing by virtue of mother tubers. But since the shoots of irradiated tubers developed strong root systems earlier than the control tubers and started feeding through them, they equalled the controls and later even surpassed them in the growth and development of their shoots. In our opinion this phenomenon is of great theoretical interest and shows the high sensitivity of potato shoots to those harmful changes which occur in the reserve nutrients of the mother tubers, under the effect of high irradiation doses of 2 c.

Determination of vitamin C content was done in six replications on the potato variety Priekul'skii during its storage from October 13 to February 17 (Table 4.25).

As evident from the data presented, the vitamin C content in the irradiated potato variety Priekul'skii was always higher than the control. The increased vitamin C content, when calculated in terms of yield, gives an additional 36 mg of ascorbic acid per 1 kg of potatoes. Considering that the preplanting irradiation of tubers increases their yield, we can also view gamma-irradiation as a measure to considerably increase the total accumulation of vitamin C in potatoes.

Table 4.25. Vitamin C content in the potato variety Priekul'skii at harvest and during its storage, mg%

Dose of irradiation, c	October 13	November 24	December 1	December 6	December 26	February 17
Control	35.3	28.8	23.8	22.1	13.7	12.5
0.5	40.7	31.4	25.6	24.4	15.0	12.9
2.0	38.9	34.1	26.4	25.2	15.8	13.8

Besides, we studied the effects of low dose irradiation on the second generation of tubers, for which potato seeds of the variety Priekul'skii grown from tubers irradiated in 1960 with 0.5 c were planted without fresh irradiation in 1961. The data thus obtained showed the presence of the stimulation effect in the second generation. Whereas, the yield obtained from the irradiated tubers was 39.4% in the year of irradiation and 23% higher than that of the control in the second generation.

In 1957 D.V. Lipsits also experimented on the irradiation of potato tubers of the varieties Kornea and Voltman. Tubers of the variety Kornea were irradiated with 0.2, 0.4 and 0.8 c. With irradiation of 0.4 and 0.8 c, he observed the stimulation of plant growth and development, which were expressed in the increase of potato yield (Table 4.26).

Table 4.26. Yield of potatoes grown from irradiated seeds

Dose of irradiation, c	Potato variety	Yield, q/ha	In percentage of the control
Control		192	100.0
0.2		249	129.7
0.4	Kornea	240	125.1
0.8		266	138.5
Control	Voltman	154	100.0
0.8		192	124.0

As evident from the table, both varieties have almost identical radiosensitivity. Increased yields with irradiation of 0.4–0.8 c varied between 24 and 38.5%.

Fischnich, Patzold and Hellinger [280] irradiated potato tubers with 0.01–0.1 c. Low radiation doses enhanced the sprouting of buds and increased their number. However, this positive effect of low dose appeared differently in different potato varieties (Akerzeger, Bona, Olympia, Vera, Corona). In this work the authors presented data that the effects of irradiation can change significantly depending on the degree of bud sprouting, temperature and the storage of irradiated tubers between irradiation and planting. According to the data from Fischnich, et al., all these factors have changed the effect of stimulation doses in such a way that it either did not appear at all or stunted the development of potato buds.

Vidal [362] in France irradiated potato tubers with ^{60}Co gamma-rays. He observed the stimulation effects of irradiation with doses of 0.5–1 c. The stimulation doses mentioned by Vidal are somewhat higher than the doses applied by researchers in the USSR. Possibly, this difference is the result of the use of different potato varieties or due to the compound effect of the soil and climatic conditions under which experiments were conducted.

Along with the acute preplanting irradiation of potato tubers, they were also subjected to chronic irradiation in the gamma-field [213]. Tubers of the potato varieties Lorkh and Priekul'skii rannyi were planted in different areas of the gamma-field of the All-Union Research Institute of Fertilizers and Agricultural Soil Science. Its source of irradiation was a cobalt isotope with an activity of 225 c. Irradiation was done for 18 hours a day; the source did not function for eight hours.

The stimulation of the growth and development of potatoes in the gamma-field appeared with the cumulative irradiation dose of 0.01–0.2 c in the vegetative period. Such a dose was ensured by planting tubers 40–150 m from the source of irradiation.

A comparison of data on the chronic and acute irradiation of potatoes shows the lesser damaging effect of gamma-rays for chronic irradiation in the gamma-field. This effect may be explained by the fact that during irradiation the tubers and sprouting buds are protected by a layer of soil in the initial, the most radiosensitive stages of germination.

Experiments on growing potatoes in the gamma-field with a 1 c source of radiation were also conducted at the Institute of Biology of Latvian SSR; the growth and development of potatoes were stimulated under conditions of chronic irradiation with the cumulative radiation dose of 0.1–0.4 c. A maximum yield increase, exceeding the control by 12–22%, appeared with doses of 0.24–0.84 c. The authors paid attention to the fact that with chronic irradiation in the gamma-field the increase of yield was attributed to enlarged tubers, whereas with acute irradiation it was a result of an increased number of tubers per plant.

To sum up the above data, it must be mentioned that the irradiation of tubers, which are the organ of vegetative propagation, stimulates the growth and development of plants. Irradiation of potato seeds shows the lesser radiosensitivity of tubers in comparison with seeds.

In the past few years experiments were conducted abroad on the preplanting irradiation of potatoes [273]. Coldera in Italy experimented with tubers of the potato variety Uaerle. The tubers were irradiated with doses of 0.1, 0.2, 0.25, 0.3, 0.35, 0.4, 0.45, 0.5 and 0.6 c on April 8; they were planted on April 24 at the rate of 32 tubers per plot with an area of 8.4 m^2. The experiments were repeated four times. Planting was random. The degree of tuber sprouting depended on the irradiation dose (0.35–0.5 c); on the basis of sprouting the tubers were divided into two batches. The number of plants, total number of tubers, their mass and diameter of tubers were taken into consideration during the harvest. Tubers of 35 mm diameter are considered marketable produce in Italy and much smaller tubers are excluded. The best method, in the opinion of the author, makes it possible to compare the treatment with a control. It is the estimate of yield on the basis of the mass of tubers per plant. Results of the experiment are shown in Table 4.27.

It is evident from the table that irradiation in all doses resulted in an increase in the mass of tubers per plant in comparison with the control, but the best results (17%) were observed with irradiation at 0.1 c, which agrees with experimental data of other radiobiologists.

In Hungary, Menyhert and Balint conducted field experiments on the preplanting irradiation of tubers of the local varieties of potato, viz., Kisvardai Roza and Gulbaba. The doses tried were 0.4, 0.6, 0.8 c with the strength of 0.01 and 0.06 c/hour. They could not obtain positive results in the first year because the irradiated tubers were stored for a long time before planting. In subsequent years of planting tubers 24 hours after irradiation, the authors observed a constantly emerging stimulation effect (Table 4.28).

Table 4.27. Results of the field experiment on the gamma-irradiation of tubers of the potato variety Uaerle (Italy)

Dose of irradiation, c	No. of plants	Total number of tubers	Tubers with diameter not less than 35 mm	Total mass, kg	Average mass of tubers per plant, kg	In percentage of the control
Control	257	2,317	1,576	253.09	0.98	100
0.10	115	1,233	943	132.44	1.15	117
0.20	127	1,259	834	133.32	1.05	107
0.25	123	1,205	836	129.25	1.05	107
0.30	102	979	751	110.93	1.09	111
0.35	85	909	537	93.65	1.10	112
0.40	122	1,266	770	129.20	1.06	108
0.45	117	1,121	809	124.61	1.06	108
0.50	94	986	594	99.78	1.06	108
0.60	125	1,201	837	128.66	1.03	105

Table 4.28. Results of field experiments on the gamma-irradiation of tubers of the potato variety Kisvardai Roza

Dose of irradiation, Krad	Number of plants	Number of tubers per plant	Total mass of tubers, g	Total mass of tubers in comparison with control, %
Control	44	26.59±2.76	1,170	100
0.5	40	28.10±3.4	1,124	96.07
0.1	42	40.45±3.85	1,689	144.36
0.3	43	39.40±3.05	1,694	144.79
0.5	45	29.31±9.21	1,319	112.74
0.7	45	31.07±7.99	1,398	119.49
1.0	42	34.38±3.2	1,444	123.42
1.5	42	32.45±3.4	1,365	116.5

The field experiment was conducted in six replications in random plots with 30 plants in each plot. Doses of 0.5–1.5 c produced the stimulation effect.

The results of preplanting irradiation of potato tubers of the variety Gulbaba are shown in Table 4.29.

As seen from the table, a 31.5% yield increase was obtained through the irradiation of tubers with 0.6 c. It must be mentioned that the amplitude of variations of the stimulation dose for the variety Kisvardai Roza was considerably wider than that for the variety Gulbaba and for varieties irradiated in the USSR, viz., Lorkh and Priekul'skii.

Apart from the tuber yield in the experiment with variety Gulbaba, the vitamin C content was also estimated (Table 4.30). The data showed that the vitamin C content increased with an increased dose.

Table 4.29. Average yield of the potato variety Gulbaba grown from tubers irradiated by gamma-rays

Dose of irradiation, Krad	Strength of radiation dose, Krad/hour	Yield from the plot, kg	In percentage of the control
Control		9.52±0.53	100.00
0.4	0.01	10.32±0.26	111.00
0.6	0.01	9.24±0.95	100.21
0.8	0.01	8.62±0.59	93.49
0.4	0.06	9.89±0.26	107.26
0.6	0.06	12.13±0.42	131.50
0.8	0.06	10.77±0.42	116.8

Table 4.30. Vitamin C content in potatoes grown from tubers irradiated by gamma-rays

Dose of irradiation, Krad	Strength of irradiation dose, Krad/hour	Vitamin C content, mg %	In percentage of the control
Control		36.2	100
0.4	0.01	43.6	120
0.6		46.1	127
0.8		51.3	141
0.4	0.06	40.3	111
0.6		44.3	122
0.8		50.7	140

Working on the irradiation of potatoes for a number of years, the authors noticed the good reproducibility of the stimulation effect and consider it possible to use this effect in national economy to get an 11–31% increased yield with a much higher vitamin C content in the tubers, surpassing the control by 11–40%.

In this way, we see the accumulation of the experimental data of several field experiments conducted in the USSR by different scientists. Field experiments on the irradiation of different potato varieties had been conducted in Italy and Hungary as well. Scientists from these countries found the expression of the stimulation effect of 0.1–0.5 c doses in the 10–40% increased yield. These doses are quite close to those used in the USSR.

In the USSR, apart from field experiments the Institute of Potato Cultivation conducted for six years, large-scale production trials on an area of about 400 ha confirmed the positive effects of radiation: 14–27% increased potato yield with much earlier ripening and increases of 1–1.5% in the starch content of tubers and 11–40% of the vitamin C content in comparison with the control.

I.I. Zeinalov, A.O. Aliev and R.R. Riza-Zade [107] in the Research Institute of AzSSR, as a result of several years experimentation, established a direct correlation between the stimulation of the potato development and its *Phytophthora* infection. The object of their research was the released commercial variety Lorkh. The tubers were irradiated three to five days before planting with doses of 0.1, 0.2, 0.3, 0.5, and 0.8 c with the strength of 740–690 c/min (Table 4.31). Optimal stimulation for potato growth and development was observed through irradiation of tubers with 0.5 c.

With an intensive *Phytophthora* infection of the haulm, the disease was reduced as a result of irradiation with the optimal stimulation dose of 0.5 c. The infection of control tubers was 34.9–52.5%. Through tuber irradiation with 0.5 c the infection was reduced up to 16.6–29.9%.

Table 4.31. Effects of the preplanting irradiation of tubers on the growth and development of potato

Dose of irradiation, c	Plant height, cm	Number of stems	Haulm mass, g	Days from planting to complete germination
Control	51.7±0.9	4.0±0.6	1,040±6	30
0.1	52.0±0.6 (−)	4.7±0 3 (−)	1,120±10 (+)	30
0.2	56.3±0.9 (+)	5.0±0.0 (+)	1.210±6 (+)	29
0.3	64.0±0.6 (+)	6.3±0.3 (+)	1,394±14 (+)	28
0.5	81.0±1.1 (+)	7.0±0.6 (+)	1,594±3 (+)	27
0.8	65.0±0.6 (+)	5.3±0.3 (+)	1,300±6 (+)	28

Note: (+) difference statistically significant, (−) difference statistically insignificant.

The experiment so conducted shows the prospects of the presowing irradiation of potato tubers with 0.5 c. The significance of irradiation increases because this dose decreases the *Phytophthora* infection of potato tubers.

Thus, the results obtained in the experiments listed in Table 4.32 enable us to recommend the preplanting irradiation of potato tubers for wider use in practical agriculture.

Sugarbeet: In the Institute of Plant Physiology of the Academy of Sciences of the UkrSSR [74], over a period of several years much research had been conducted on the presowing irradiation of air dried sugarbeet seeds. The researchers have noticed an increase in the yield and sugar content of tubers grown from irradiated seeds. In 1960 in commercial sugarbeet crops grown from seeds irradiated with 0.5, 1 and 5 c, the highest yield of tubers was obtained from crops raised from seeds irradiated with 1 c. A dose of 5 c insignificantly affected the increased yield potential of tubers, but the sugar content increased by 2.2% (Man'kovskii region), which gave an additional yield of about ten quintals of sugar from one hectare (Table 4.33).

In field experiments conducted on sugarbeet under conditions of the experimental base of the Institute of Plant Physiology of the Academy of Sciences of UkrSSR, there was a 21–26 q/ha increase in the yield of tubers over the yield of control plants, which was 386 q/ha. Here, the sugar content increased by 0.7–0.9%. The authors explain the increase in the yield of sugarbeet by the activation of the physiological processes controlling the stimulation effects of irradiation. For example, irradiation increases the photochemical activity of chloroplasts and changes the activity of oxidizing enzymes like peroxidase, polyphenoloxidase and catalase. A spectrographic analysis of the ash, showed an increased selective capability of plants, as a result of irradiation, regarding micro and ultramicronutrients, which, in the opinion of P.A. Vlasyuk, leads to an increase of plant productivity and the

Table 4.32. Consolidated data on the preplanting irradiation of potato tubers

Place of research	Year	Researcher	Dose of irradiation	Increase of yield in comparison with control, %	Change of biochemical composition	Enhancement of development
Institute of Potato Cultivation	1959	A.I. Grechushnikov	0.3 c	Variety Lorkh 8–19 Variety Priekul'skii 12–44	—	Enhancement of bud sprouting
Institute of Biology of the Academy of Sciences of Latvian SSR	1959	K.K. Roze, G.E. Kavatse	Chronic 0.24–0.48 c	12–22	— —	Enhancement of bud sprouting
Institute of Biophysics of the USSR Academy of Sciences	1961	N.M. Berezina, R.R. Riza-Zade, G.I. Shchibrya	0.5 c	38	Increase of vitamin C content	Enhancement of bud sprouting
Institute of Potato Cultivation	1960	V.S. Serebrenikov	0.3–0.5 c	22–36 20–44	Increase in starch content	—
State farm Kolomenskii	1962	V.S. Serebrenikov	0.5 c	17–18	—	—
Experimental farm Orlovskoe	1962	V.S. Serebrenikov	0.5 c	16–18	—	—
State farm Kolomenskii	1963	V.S. Serebrenikov	0.5 c	18.9	—	—

Collective farm Put' Novoi Zhizni	1963	V.S. Serebrenikov	0.5 c	17–18	—	—
Italy (Camo)	1968	Calder	0.1 Krad	17	—	—
Hungary	1967	Menyhert	0.6 c 0.8 c	31.1	Increase in the vitamin C content by 11–41%	—
Hungary	1968	Balint	0.1 c 0.3 c	44.4 44.7	—	—

Table 4.33. The effect of ^{60}Co gamma-irradiation of sugarbeet seeds on the yield and sugar content of tubers (data of 1960)

Dose of irradiation, c	Yield, q/ha	Increase of yield in comparison with the control, q/ha	Sugar content, %
Collective farm named after 300 years of reunion of the Ukraine with Russia, Vitebskii region of Ternopol' province			
Control	237	—	19.7
0.5	250	13	21.0
1.0	248	11	20.1
5.0	244	7	19.6
Collective farm of the Man'kovskii region of Cherkass province			
Control	254	—	17.7
0.5	275	21	16.8
1.0	295	41	19.5
5.0	276	22	19.9
T.G. Shevchenko collective farm of the Prilukskii region of Chernigov province			
Control	400.0	—	—
0.5	418.5	81.5	—
1.0	433.2	33.2	—
5.0	401.2	1.2	—
Collective farm Kommunist of the Rovenskii region of Sumskaya province			
Control	313.2	—	—
0.5	320.0	6.8	—
1.0	309.3	3.9	—
5.0	306.1	7.1	—

improvement of the quality of agricultural produce. The authors make this conclusion on the basis of 240 experiments conducted during nine years in different soil-climatic zones of the Ukrainian SSR on sugarbeet and other agricultural crops. In most of the experiments there was a 10–20% increase in the yield. Only a few experiments showed negative results: the yield was reduced by 5% or was no different from the yield of the control plants. These data served as a basis to conduct trials on the presowing irradiation of seeds under large-scale production conditions.

In 1962 such a trial was conducted at 16 places under Gossortoset', the state trial network. In some regions the yield increased through the irradiation of seeds with the dose of 0.5 c, while in others, through irradiation in the dose of 1 c. At two centers, neither of these doses produced any stimulation. Such variations in the results show the inadequacy of our knowledge concerning the compound effects of the soil-climatic conditions on the effects of irradiation.

In the L'vov State University, during many years since 1957, experiments were conducted on the presowing irradiation of sprouted sugarbeet seeds of the varieties Lebedinskaya, Ramonskaya 931 and Odnorostkovaya with doses of 0.5 and 1 c and sowing them in sandy loam and loam soils (Table 4.34).

As evident from the data from small plot experiments presented in the table, the presowing irradiation of sprouted sugarbeet seeds is more effective than the presowing irradiation of air dried seeds, but it produces positive effect only with regard to the tuber yield. The sugar content which was, however, not always determined by the researchers, was reduced everywhere.

At Gorky University, in experiments on the presowing irradiation of air dried sugarbeet seeds, apart from yield and sugar content, the activity of respiratory enzymes was also taken into consideration (Table 4.35) [80].

The data obtained show very clearly that the presowing irradiation of seeds produces a very distinct effect of distant action in the form of increased enzymatic activity of leaves, particularly during early stages of growth and development of the plants. In these experiments, the presowing irradiation did not influence the yield of tubers, but increased their sugar content. As a result the sugar output (quintals) per hectare was higher than in the control, particularly with the irradiation of 0.5 c (Table 4.36).

Field experiments on the presowing irradiation of sugarbeet seeds were conducted for several years by A.T. Miller [168, 170] at the Institute of Biology of the Academy of Sciences of the Latvian SSR. Sugarbeets cultivated in the northern regions, including the Latvian SSR, as is well known, have low sugar content. Therefore, the objective of these experiments was to increase the sugar content of this crop, possibly through the gamma-irradiation of seeds.

Air dried seeds of sugarbeet varieties MO-70, MO-80 and Yaltushkovskaya odnosemennaya, which have been released for cultivation in the Latvian SSR, were irradiated 10–20 days before sowing in the radiation flux of the nuclear reactor with 2 c with the strength of 10 c/sec and the radiation energy of 1.5 MeV. In 1962, through the irradiation of seeds of the variety Yaltushkovskaya odnosemennaya with 2 c, the tuber yield was 40.3 q/ha with a sugar content of 16.9%, whereas the yield of unirradiated tubers was 36.9 q/ha with a sugar content of 16.38%. The field experiments were repeated under commercial cultivation conditions. In the years favorable for sugarbeet the increase in yield was much higher. The results of commercial trial production during 1963–1965 are presented in Tables 4.37 and 4.38.

In one of the collective farms where large-scale production trials on the presowing irradiation of sugarbeet seeds were conducted, the commercial viability of this measure was also estimated. The sugar output in three years

Table 4.34. Effect of the presowing irradiation of sprouted seeds on the yield and sugar content of sugarbeet

Dose of irradiation, c	Year	Variety	Soil	Yield, q/ha	In percentage of the control	Sugar content, %
Control	1957	Lebedinskaya	Sandy loam	119±24	100	17.2
0.5				150±15	126	14.7
1.0				158±10	133	13.4
Control	1958	Ramonskaya 931	Sandy loam	267±40	100	17.7
0.5				330±50	125	17.4
1.0				416±57	156	16.9
Control	1960	Lebedinskaya	Loam	246±32	413	17.4
0.5				300±41	146	16.9
1.0				360±43	100	—
Control	1960	Ramonskaya 931	Loam	341±32	100	—
0.5				—	—	—
1.0				541±35	157	—
Control	1960	Odnorostkovaya	Loam	337±32	—	—
1.0				538±31	142	—

Table 4.35. Activity of respiratory enzymes in sugarbeet leaves (mg of ascorbic acid oxidized by the enzyme in 30 min. per 1 g of dry matter)

Dose of irradiation, c	Peroxidase			Polyphenoloxidase			Ascorbinoxidase		
	June 13	June 20	July 30	June 13	June 20	July 30	June 13	June 20	July 30
Control	150.7	110.9	133.8	39.4	35.0	60.2	2.8	2.0	3.1
0.5	365.0	187.0	187.4	51.9	55.2	130.7	3.3	4.2	5.7
1.0	373.2	196.1	129.8	112.1	47.0	77.0	3.3	5.2	6.2

Table 4.36. Effect of the irradiation of sugarbeet seeds on the yield and sugar content of tubers

Dose of irradiation, c	Yield, q/ha	Sugar content, %
Control	207	16.0
0.5	204	19.1
1.0	204	18.6

Table 4.37. Results of the large-scale production experiments on the presowing irradiation of sugarbeet seeds with 2 c in the Latvian SSR

Year	Area, ha	Yield, q/ha		Sugar content, %	
		Control	Irradiation	Control	Irradiation
1963	4.8	232.3±8.4	246.4±9.7	15.6±0.06	15.8±0.10
1964	126.6	322.2±7.0	334.4±7.6	16.7±0.1	17.2±0.07
1965	180.0	285.8±5.3	295.5±9.8	16.3±0.074	16.7±0.075
Average		280.1±4.05	292.1±5.2	16.2±0.046	16.57±.0.047

Table 4.38. Output of sugar per hectare with the presowing irradiation of sugarbeet seeds with 2 c

Year	Area, ha	Output of sugar, q/ha		Additional output of sugar, q/ha
		Control	Irradiation	
1963	4.8	36.15	38.86	2.71
1964	126.6	54.50	58.13	3.63
1965	180.0	46.47	49.52	2.76
Average		45.70	48.83	3.04

of trials was 3 q/ha higher than in the control. During the evaluation of the produce for production cost, it was established that the net profit per hectare was 90 rubles with the expenditure for irradiation only 1.5 ruble per hectare. When these trials were continued to 1968, the average increase in the output of sugar for seven years was 4.04 q/ha. The yield as well as the sugar output per hectare increased stably over a period of seven years. Trials on the presowing irradiation in the Gossortoset', the state varietal trial program, showed an additional output of sugar to the tune of 4.4 q/ha for three years.

An assessment of the presowing irradiation of sugarbeet seeds in field experiments and large-scale production trials for many years, enables us to consider it a possible measure for use in the practical agriculture in the Latvian SSR.

In Czechoslovakia, Drachowska and Sandero [278] conducted experiments on the presowing irradiation of sugarbeet seeds by ^{60}Co gamma-rays to increase the sugar content and reduce its loss during the storage of marketable tubers. According to their data, irradiation with 1 c influenced the increase of yield more than the increase of sugar content. The output of sugar from one hectare was 15–24% more in comparison with the control.

According to data from the Hungarian scientist Pannonhalmi [334], through the irradiation of sugarbeet seeds with 1–5 c there was increase in the yield of tubers as well as their sugar content.

Research workers at the L'vov State University conducted the presowing irradiation of sprouted seeds of fodder beet varieties Ideal and Poltavskaya as well as seeds of the table purpose best variety Egipetskaya (Table 4.39).

The irradiation of sprouted seeds of fodder and table purpose beet varieties with 0.5–3 c led to a 20–40% increase in the yield of tubers in comparison with the control.

Table 4.39. Effects of the presowing irradiation of sprouted seeds of fodder and table purpose beet varieties on the yield of tubers

Dose of irradiation, c	Year	Beet variety	Soil	Yield per plot, kg	In percentage to the control
Control				200±30	100
0.5	1958	Egipetskaya	Sandy loam	240±39	120
1.0				280±32	140
3.0				240±32	120
Control				450±49	100
0.5	1959	Ideal	Sandy loam	501±58	123
1.0		Poltavskaya		534±79	132

P.G. Agakii and K.I. Sukach [3] studied the effects of the presowing irradiation of sugarbeet seeds at different nutrition levels under the conditions of the MSSR. The seeds were irradiated with doses of 0.5, 1, 1.5, 2, 2.5 and 3 c and stimulation occurred in all the doses except the last. The optimal stimulation doses were 2 and 2.5 c, as irradiation in these doses led to a 0.6–1.2% increase in the sugar content of tubers. In 1972, they conducted large-scale production trials on the irradiation of seeds of this crop on an area of 100 ha.

In the All-Russian Research Institute of Sugarbeet and Sugar [204], experiments were conducted on the presowing irradiation of the sugarbeet varieties R-100 and R-09 with 1, 5 and 10 c with 10–12% moisture content in the seeds. The optimal stimulation dose in these experiments was 1 c, which made it possible to increase tuber yield by 15–18.7 q/ha. The irradiation of seeds with 1 c facilitated the increase of good quality juice and syrup diffused in both varieties. In this dose, the sugar content of molasses was reduced, which ultimately increased the output of sugar by 0.55% for multi-seeded beets and by 1.7% for single-seeded beets on the basis of mass of the tubers in the sugar factory. In this way, the irradiation with 1 c, as per the data from the institute, can be considered an effective measure for increasing the yield and improving the technological properties of sugarbeet tubers.

While irradiating sugarbeet seeds in a wide range of doses, V.I. Kostin [139–141] established that the optimal stimulation doses are 1–1.2 for Ul'yanove province. According to his data, the tuber yield in field experiments increased by 16% for four years and by 11% under large-scale production conditions. The output of sugar per hectare, under large-scale production conditions, increased by 6.3 q/ha. He also noticed the continuous nature of the stimulation: the irradiated plants entered every phase of development earlier and also completed it earlier.

Table 4.40 shows the results of the presowing irradiation of sugar, fodder and table purpose beet varieties.

Tomato. A large-scale experiment on the irradiation of tomato seeds of the variety Luchshii iz Vsekh (Best of all) in doses of 0.3, 0.5 and 1 c by growing them in greenhouses was conducted at Gorky University [6]. The experiment showed that the optimal doses were 0.3 and 0.5 c (Table 4.41). Analogous work was conducted by N.P. Chernov [245].

The economic effectiveness of irradiation was calculated: the total sum obtained through the sale proceeds of irradiated tomatoes in terms of output per 1 m^2 exceeded the sum obtained from the sale proceeds of control tomatoes by 1.3–4.4 rubles. This increased the profit from the sale proceeds of produce from irradiated seeds by 10.3–28% (Table 4.42).

Sprouted seeds of the tomato variety Donetskii were irradiated with 0.5 and 1 c in the L'vov State University. As per the data from the yield estimate of 1960, 900 g of fruits were harvested per plot in the control, on an average, while irradiation with 0.5 and 1 c produced fruit yields of 1,200 and 1,500 g respectively, which amounted to an increase by 33 and 66% in comparison with the control [78].

Three-year tests on the presowing irradiation of the tomato variety Luchshii iz vsekh were completed in 1970. The seeds were irradiated with 0.25, 0.5, 1, 2 and 4 c on the GUPOS unit at the Institute of Biophysics of the USSR Academy of Sciences. The strength of the dose was 700 r/min [1, 30, 31]. Tomatoes were grown under hydroponics in the greenhouses of the petroleum refinery in Moscow province. The criteria for evaluating the stimulation effect of gamma-irradiation were plant height, size of the assimilation surface, fruit yield and fruit quality. The optimal stimulation dose under hydroponic conditions was 2 c (Table 4.43).

The optimal dose for growing tomato under hydroponics (2 c) was considerably higher than the stimulation doses when growing them in the field (0.5 c) and in greenhouses (1 c). We presume that the reason for the increase in the stimulation doses for growing tomatoes in hydroponics is the washing

Table 4.40. Consolidated data on the presowing irradiation of seeds of sugar, fodder and table purpose beet varieties

Place of research	Year	Researcher	Dose of irradiation, c	Increase of the yield of tubers in comparison with the control	Sugar content
Institute of Plant Physiology of the Academy of Sciences of the UkrSSR	1953–1962	P.A. Vlasyuk, A.V. Manorik, D.M. Grodzinskii	0.5 1 5	21–26 q/ha	Increase by 0.7–2.2%
L'vov State University	1957	S.O. Grebinskii, V.G. Tsibukh	0.5–1 (soaked seeds)	26–33%	Decrease from 17.2 to 13.4%
L'vov State University	1958	S.O. Grebinskii, V.G. Tsibukh	0.5–1 (soaked seeds)	25–56%	Decrease from 17.7 to 16.9%
Agrophysics Institute	1958	N.F. Bateegin	1	42%	Increase by 16.5%
Agrophysics Institute	1959	N.F. Bateegin	0.5	19%	—
L'vov State University	1959 (fodder beet varieties Ideal, Poltavskaya)	S.D. Grebinskii, V.G. Tsibukh	0.5–1 (sprouted seeds)	23–32%	—
Agrophysics Institute	1959–1966	N.F. Bateegin	2–6	30 q/ha 35–40 q/ha	—
L'vov State University	1960	S.O. Grebinskii, V.G. Tsibukh	1	57	No increase

Collective farm named after 300 years of the reunion with Russia of the Vitebskii region of Ternopol' province of the Ukraine	1960	P.A. Vlasyuk	0.5	13 q/ha	Increase by 1.7%
Collective farm of the Man'kovskii region of Cherkass province	1966	P.A. Vlasyuk	1–5	41 q/ha	Increase by 2.2%
T.G. Shevchenko collective farm of the Prilukskii region of Chernigov province	1960	P.A. Vlasyuk	1	33.2 q/ha	—
Collective farm Kommunist, Czechoslovakia	1960–1961	Drakhowska	1	6.8 q/ha	Increase of output by 15–24%
Institute of Experimental Biology of the Academy of Sciences of the Latvian SSR	1963	A.T. Miller	2	14 q/ha	Increase by 0.2%
	1964		2	14 q/ha	Increase by 0.5%
	1965		2	7 q/ha	Increase by 0.4%
	1963–1965 (average)		2	12 q/ha	Increase by 0.37%

Table 4.41. Yield of tomatoes grown in commercial greenhouses of the Gorkovskii state farm from seeds irradiated by gamma-rays

Time of harvesting the fruits	1955			1967	
	Control	0.3 c	0.5 c	Control	0.5 c
	kg/m²	%		kg/m²	%
May	—	—	—	2.25	123.6
June	2.54	130.7	125.2	9.90	110.4
July	5.06	115.4	108.9	1.28	112.5

Table 4.42. Profit from the proceeds of tomatoes grown from irradiated seeds, in terms of rubles per square meter

Time of harvesting of fruits	1965			1967	
	Control	0.3 c	0.5 c	Control	0.5 c
		State farm Gor'kovskii			
May	—	—	—	5.6	7.0
June	6.3	8.3	7.9	15.1	18.2
July	6.6	7.6	7.2	2.9	2.8
Total	12.9	15.9	15.1	23.6	28.0
		State farm Zhdanovskii			
June	7.8	9.5	9.4	7.0	10.0
July	4.2	3.8	3.9	3.7	3.7
Total	12.0	13.3	13.3	10.7	13.7

Table 4.43. Effects of the gamma-irradiation of tomato seeds on the fruit yield

Dose of irradiation, c	Yield from the experimental area, kg	In percentage of the control	Criterion of significance
	1965		
Control	7.3±0.15	100.0	2.4
0.5	8.1±0.17	110.9	4.0
1.0	8.2±0.15	112.3	4.2
2.0	8.6±0.15	117.3	5.1
	1966		
Control	7.8±0.24	100.0	—
2.0	9.4±0.10	120.5	3.1
4.0	8.0±0.17	102.5	0.3
10.0	7.0±0.17	89.7	1.3

away of radiotoxins from the roots of the irradiated plants as a result of excessive irrigation. The work of V.A. Kopylov and A.M. Kuzin [138] and A.M. Kuzin and Yu.N. Runova [147] also testifies to the possible washing away of radiotoxins.

This experiment also showed the useful biochemical changes that emerged in tomatoes grown from seeds irradiated in the stimulation dose; particularly the 22.5–29.7% increase in the ascorbic acid content and of carotine by 27.3–29.8% (Table 4.44).

For a period of three years I.G. Suleimanova [221] irradiated tomato seeds of the variety Mayak 12/20 with doses of 2, 4, 8 and 16 c with the strength of 720 r/min. The stimulation effect was observed in a wide range of doses—from 2 to 10 c. The optimal effect was manifest with the irradiation of seeds grown on unfertilized soil at 4 c, but at 8 c on the fertilized N_{200} P_{200}) soil. Therefore, with the high level of mineral nutrition of the soil, the maximum manifestation of the stimulation effect is possible through a much higher dose of irradiation. The increase in fruit yield as a result of the irradiation of seeds at 8 c was 11.2 and 14% in different years. The stimulation doses significantly changed the content of dry matter in the fruits, which was at its maximum with irradiation at 16 c.

Presowing irradiation of tomato seeds was conducted by Kersten [308]. According to data from the Canadian researcher MacKey, the fruit yield increased by 21–27% with much earlier ripening through the irradiation of seeds with 0.3 c. The experiment was conducted for two years, 1969–70.

During 1964–1967, presowing irradiation of the best Hungarian varieties of tomato was done in Hungary. Of significant interest is the research from the Institute of Agricultural Radiobiology. They irradiated seeds of mature fruit with ^{60}Co gamma-radiation and later washed, dried and sowed them. They tried irradiation doses of 0.1, 0.2, 0.3 and 0.5 Krad. The experiments were conducted under greenhouse and field conditions [323, 324].

At the University at Grodollo air dried seeds of the Hungarian tomato variety Budei-Korat were irradiated with 0.5, 0.75, 5 and 10 Krad. The experiments were conducted simultaneously in a phytotron and in small field plots. Under controlled conditions, the plants were grown in water culture in the Knopp mixture. Illumination was given for 12 hours a day at the intensity of 10 lux with fluorescent lamps and the air humidity varied from 55 to 60%. The field experiments were conducted in four replications on random plots with an area of 40 × 41 cm. Besides the total yield, data on the periods of flowering, total number of flowers and number and mass of fruits were also taken into consideration.

The amount of carbohydrates in the leaves varied during the vegetative period: by the seventh week of vegetative growth, it comprised 23–49.5%, then reduced to 5–20%. At the time of flowering the amount of carbohydrates again increased. In the leaves of plants irradiated with 1–5 c, the carbo-

Table 4.44. Effects of gamma-irradiation of tomato seeds on the biochemical composition of fruits grown from them under hydroponic conditions

Dose of irradiation, c	Ascorbic acid		Carotine		Lycopin		Sugar	
	mg%	%	mg%	%	mg %	%	mg%	%
			1965					
Control	37	100.0	1.10	100.0	6.6	100.0	2.5	100.0
0.5	43	116.2	1.10	100.0	6.7	101.5	2.7	108.0
1.0	46	124.3	1.10	100.0	6.8	103.0	3.0	120.0
2.0	48	129.7	1.40	127.3	7.4	112.1	3.4	136.0
			1966					
Control	31	100.0	1.14	100.0	6.15	100.0	1.6	100.0
2.0	38	122.5	1.48	129.8	6.74	109.5	2.0	125.0
4.0	35	112.9	1.42	124.5	6.39	103.9	1.8	112.5
10.0	31	100.0	1.36	110.5	6.29	102.2	1.6	100.0

hydrates were 40–43% more than in the leaves of unirradiated plants. The initial growth was stunted at 10 c. The stimulation at the dose of 1 c was expressed by the increased number of leaves and flowers. The maximum number of flowers was observed with the dose of 0.75 Krad, when it comprised 170.13% of the control; at 5 Krad, it comprised 163.4% and with 10 Krad, just 36.19% of the control. The fruit yield with 0.75 Krad was as high as 108.47% and 114.12% at 1 Krad as compared with the control. Irradiated tomato seeds had fruit which ripened five days earlier when compared with the control. The data showed that the presowing irradiation of tomato seeds is a profitable measure in Hungary.

At the Congress in Milan in October, 1968, some Hungarian scientists presented an interesting report on the crossing of two botanical varieties of tomato, whose seeds were irradiated before sowing. The effect of heterosis was intensified. Their report is also of great interest in that, if the tomato pollen is irradiated with 0.4 Krad before crossing, the pollination efficiency increases and a very large number of seeds form in the hybrid fruits.

The work of German scientists also deserves our attention [347]. The authors irradiated the seeds of the tomato variety Professor Rudolph by x-rays in high doses (20, 30, 40 c), which stunted the plants grown from irradiated seeds. But stimulation appeared in the second generation of these seeds. The increase in the raw mass of second generation seedlings and that of the adult plants is presented in Table 4.45.

It is evident from the data presented that the second generation of plants grown from the seeds irradiated in high doses shows a high and stable increase in the raw mass of seedlings and adult plants.

Consolidated data on the presowing irradiation of tomato are given in Table 4.46.

Table 4.45. Raw mass of seedlings and plants of the tomato variety Professor Rudolph

Progeny	Number	Average raw mass, g	In percentage of the control
		Seedlings, 1970	
M_1	167	11.35±0.32	100
M_2	163	19.07±0.45	168
		Seedlings, 1971	
M_1	83	13.48±0.3	100
M_2	64	20.41±0.5	151
		Plants	
M_1	131	434±8	100
M_2	131	544±8	125

Table 4.46. Consolidated data on the presowing irradiation of tomato seeds

Place of research	Year	Researcher	Dose of irradiation	Increase of yield in comparison with the control, %	Change of biochemical composition	Enhancement of development
L'vov State University	1960	S.O. Grebinskii, V.G. Tsibukh	0.5–1.0 c (sprouted seeds)	33–90	—	Enhanced ripening
Gorky University	1970	I.F. Aleksandrova	0.5–1.0 c	10–12	Increased content of ascorbic acid	Enhanced ripening
Institute of Biophysics of the USSR Academy of Sciences	1970	M.A. Abdulaev	2 c	12–20	Increased content of ascorbic acid by 22–29 mg%, carotine by 27–29 mg%, lycopin by 9–12 mg%	Enhanced ripening
Canada	1968–1971	MacKey	0.5–1 Krad	34	—	Much earlier ripening
Hungary	1964–1968	Pal, Simon, Baldi	0.75 Krad 1 Krad	8.47 14.12	—	Enhanced ripening by 5 days

Lettuce: Evler [279] was one of the first researchers to note that the generation and further development of lettuce accelerates as a result of the irradiation of its seeds.

According to the data of S.O. Grebinskii and V.G. Tsibukh [78], the maximum increase in the yield of green mass of lettuce was obtained through the irradiation of sprouted seeds with 0.25 c, but 1 c stunted the growth and development of lettuce to 76% in comparison with the control.

M.A. Abdulaev conducted presowing irradiation of lettuce seeds with 0.25, 0.5 and 1 c. The optimal dose was found to be 0.25 c, in which the yield of green mass of the plants increased by 18%. Simultaneously the quality of lettuce also improved as a result of the 8 mg% increased ascorbic acid content in the leaves in comparison with the control. The author also observed the increased chlorophyll and carotinoids content in the leaves of irradiated plants [30, 32].

In 1968 the presowing irradiation of lettuce seeds was done in Canada with doses of 0.1–1 c. Here, according to the data from MacKey, there was a 30–50% increase in the yield and a much earlier ripening of the green mass.

The results of experiments on the irradiation of lettuce seeds are shown in Table 4.47.

Muskmelon, watermelon and pumpkin: The first reports on the irradiation of the seeds of cucurbitaceous plants were made by Evler [279]. According to his data, irradiation facilitated an increase in the germination energy of seeds and enhanced the growth and development of the plants. In the USSR experiments on the presowing irradiation of cucurbitaceous plants were conducted at the Institute of Genetics and Plant Physiology of the Academy of Sciences of the AzSSR. The cultivation of cucurbitaceous plants as fodder and food crops is of great national economic importance for Azerbaijan. The application of presowing irradiation of the seeds of these crops is promising because, according to data in the literature, irradiation can accelerate plant ripening, increase their yield potential and increase their sugar and vitamin contents (each of these changes has great economic importance in the cultivation of cucurbitaceous crops). The experiments were conducted at the experimental base of the Institute of Genetics and Plant Physiology of the Academy of Sciences of the AzSSR situated on the Apsheron Peninsula. Seeds of the watermelon varieties Melitopol'skii and round, black Kakhetinskii and those of the pumpkin varieties Stofuntovaya and Karotinnaya were irradiated by ^{60}Co gamma-rays with the doses of 10, 20, 30 and 40 ç and a strength of 560 r/min. The irradiation of seeds of the muskmelon varieties Kolkhoznitsa and Gurbek was done with 5, 10, 20, 30, and 40 c. Such high doses of irradiation were taken on the basis of the laboratory experiments which showed the very high radiosensitivity of the representatives of the Cucurbitaceae family [239]. The maximum increase in the

Table 4.47. Consolidated data on the presowing irradiation of lettuce seeds

Place of research	Year	Researcher	Dose of irradiation	Increase of yield in comparison with the control, %	Change of biochemical composition	Enhancement of development
L'vov State University	1957	S.O. Grebinskii, V.G. Tsibukh	x-radiation, 0.25 c (sprouted seeds)	65	—	Enhancement of ripening
Institute of Biophysics of the USSR Academy of Sciences	1970	M.A. Abdulaev	0.25 c	20	Increase in the content of ascorbic acid	—
Canada	1968–1970	MacKey	0.1–1.0 Krad	30–50	—	Much earlier ripening

muskmelon yield was noticed with the irradiation of seeds at 5 c, and at 10–20 c for watermelon. In all cases of stimulation known to date due to the effects of gamma-irradiation, this is the first time we have come across such a high dose as 20 c. Table 4.48 shows the data of the field experiments.

Table 4.48. Yields of muskmelon and watermelon raised from air dried, irradiated seeds

Dose of irradiation, c	Muskmelon variety Kolkhoznitsa		Watermelon variety Melitopol'skii	
	Yield, q/ha	In percentage of the control	Yield, q/ha	In percentage of the control
Control	198.8	100.0	44.2	100.0
5	227.3	115.0	—	—
10	207.6	105.0	44.1	100.0
15	—	—	—	—
20	100.0	50.5	50.2	113.5
25	83.2	41.9	—	—
30	57.8	29.9	41.5	94

In the next year these results were verified by large-scale production conditions, simultaneously repeating the same irradiation treatments in field experiments. Muskmelon seeds irradiated with 4 c were sown in an area of 0.5 ha in the V.I. Lenin collective farm of the Astaurinskii region. The yield harvested was 248 q/ha. The yield increase in comparison with the control was about 13% (yield of the control plants—220 q/ha). Fruits grown from the irradiated seeds matured five to six days earlier than those from unirradiated seeds. As a result the collective farm had produce ready earlier than all other neighboring farms. It must be mentioned that the result of this experiment was more interesting to collective farmers than the yield increase caused by presowing irradiation, from the viewpoint of economic profit.

Phenological observations show that the maximum enhancement of plant development occurs in the initial stages (from seedling emergence to flowering). The repeated, large-scale commercial sowing of muskmelon and watermelon seeds in the same farm confirmed the above data. It is worthwhile to note that through seed irradiation in doses that cause stimulation, plants had more intensive whisk formation. If an average of three or four stolons were seen on the control plants, then up to nine stolons developed in plants grown from seeds which were irradiated with 4 c. Earlier we considered this phenomenon a result of awakening additional buds by stimulation doses of irradiation.

Along with morphological and physiological changes, biochemical changes were also observed in the fruits: through the irradiation of seeds with 16 c the soluble sugar content increased in muskmelon and watermelon, but decreased in pumpkin. The maximum amount of sugar in muskmelon and pumpkin was obtained when their seeds were irradiated with 4 c. All irradiation doses, even those which slightly suppressed the growth and development of plants, increased the sugar content of muskmelon and watermelon over the control.

The vitamin C content was estimated in these varieties of muskmelon, watermelon and pumpkin by Til'men's method.

With 4 c there appeared a very significant increase of Vitamin C—up to 23.5 mg% in the muskmelon variety Kolkhoznitsa and 5 mg% in the muskmelon variety Gurbek. The increase in vitamin C content was considerably less in watermelon and pumpkin. In the pumpkin varieties Stofuntovaya and Karotinnaya, the carotine content increased 3.7 and 4.8 mg% respectively with the dose of 4 c. Considering that sugar and vitamin contents are basic indices of the quality of muskmelon, watermelon and pumpkin, we may confidently say that the biochemical chages in the fruits of cucurbitaceous plants as a result of presowing irradiation improved the quality of produce and increased its fodder value.

The determination of the activity of the oxidizing enzymes showed some reduction in the activity of peroxidase, which remained at the same level throughout the vegetative period. This reduced peroxidase activity may be one of the reasons for the increased ascorbic acid content in the fruits grown from irradiated seeds. The activity of catalase increased only at the initial stages of seed germination, but then it was somewhat lower than in the control plants throughout the vegetative period. Much slower rates of oxidizing processes may be a reason for the increased sugar content, which occurs quite often in the most diverse plants due to the presowing irradiation.

A list of the changes in cucurbitaceous plants due to irradiation shows how complicated and complex are the changes caused by the stimulation effect of ionizing radiation and how primitive were the concepts of early radiobiologists who thought that the stimulation effect changes only in accordance with the dose and can be determined from any one index, like length of roots, plant height, etc.

In Bulgaria Prof. Panasov conducted experiments on the presowing irradiation of watermelon seeds. The stimulation effects of irradiation in his experiments, as well as in our experiments, emerged for high doses of irradiation, i.e., 10 c. The yield of fruits grown from irradiated seeds exceeded the control by 18% and the sugar content in fruits by over 8% (cited from Glubrecht [288]).

Consolidated data on the presowing irradiation of the seeds of cucurbitaceous plants are shown in Table 4.49.

Table 4.49. Consolidated data on the presowing irradiation of seeds of cucurbitaceous plants

Place of research	Year, crop	Dose of irradiation, c	Researcher	Increase of yield in comparison with the control, %	Change of biochemical composition	Enhancement of ripening, days
Institute of Genetics and Plant Physiology of the USSR Academy of Sciences	1950 muskmelon	4	A.I. Khudadatov	15	Increase in the content of sugar by 2.3%, ascorbic acid by 5.1 mg%	3
	watermelon	20		13–15	Increase in the content of sugar by 1.23%, ascorbic acid by 1.7 mg%	7
V.I. Lenin collective farm of Astaurinskii region	1960 muskmelon	4	A.I. Khudadatov	14	—	5
	pumpkin	4		12	Increase in the content of sugar by 1.07%, ascorbic acid by 0.65 mg%, carotine by 4.8 mg%	5
Bulgaria	1969 watermelon	—	Panasov	18	Increase in sugar content by 8%	—

LEGUMES AND CEREALS

Peas. Much research on the presowing irradiation of air dried and soaked seeds was conducted on this crop. Schmidt was the first researcher who, while irradiating air dried seeds in 1910, observed the enhanced growth and development of plants due to the effects of radiation at 1/10, 1/4, 1/2 and 1 NED (NED—strength of exposition dose of gamma-rays at a distance of 1 m, r/sec). As per the data from the author, the pea plants grown from irradiated seeds developed pods more vigorously and each pod had four to five seeds, as compared to two or three seeds in the control. Schmidt's experiments led to the confirmation of doses which stimulated the growth and development of plants, since at that time, it was not absolutely clear and was controversial. Initially, Schmidt expressed the opinion that ionizing radiation must have a practical importance for floriculture and horticulture. Czepa [274] also conducted experiments on the presowing irradiation of seeds of a number of agricultural crops including pea. On the contrary, he negated the existence of the stimulation effects of low doses of radiation. These contradictions by these scientists are now understood. They did not pay attention to the variety of pea, did not have the means to measure the exact radiation doses and had no knowledge of the environmental factors and physiological conditions of the objects being irradiated, which can significantly change the effect of this or that irradiation dose for the crop under study.

Much later, research was conducted by the Soviet scientists, L.P. Breslavets and A.I. Atabekova [56] who irradiated pea seeds of the variety Ideal with doses of 0.05, 0.1, 0.2, 0.25, 0.35, 0.45, 0.55, 0.65 and 0.75 c. The irradiated seeds were grown in experimental pots. Seeds irradiated with 0.35 c germinated earlier and the plants developed faster and considerably more vigorously than the control plants and produced a maximum number of pods. After establishing the optimal dose of irradiation, the authors in 1935 planned the first field experiment on farms of the K.A. Timiryazev Academy of Agricultural Sciences and confirmed the importance of the optimal dose (0.35 c) in the presowing irradiation of the pea variety Ideal. In the experiments on the presowing irradiation of pea conducted at the Ural branch of the USSR Academy of Sciences [189] the effect of α, β, γ and x-rays was compared. It was established that the stimulation effect is produced only by gamma- and x-rays with low linear intensity of ionization. Even very low doses of alpha-particle irradiation stunted the growth and development of the pea.

A comparison was made of the optimal stimulation doses of irradiation for dry and soaked pea seeds of the variety Ideal. For dry seeds, the optimal doses were 0.3–0.5 c and 0.076–0.081 c for the soaked seeds. To clarify the role of light in the effects of gamma-irradiation, the seeds of pea irradiated with 0.25–6 c were germinated in light and in darkness. The stimulation effect of the low dose of irradiation was distinctly manifest during the development of plants in light. Growth was less stunted by high doses in light than in darkness.

A.I. Atabekova [11–13] studied the effects of x-rays on dry and soaked pea seeds of the variety Chudo Ameriki (American Wonder). Seeds were soaked for six hours, then dried and irradiated with doses of 0.05, 0.1, 0.2, 0.25, 0.35, 0.45, 0.55, 0.65, 0.75 and 1 c with x-rays at the voltage of 18 kV and 4 mA current, using a 0.5 mm thick aluminum filter. The soaked seeds were irradiated with doses of 0.05–0.55 c at the same intervals. A cytological analysis of the seedlings and dry seeds showed an absence of any deviations in the development of plants with irradiation at 0.05 and 0.1 c. Beginning with 0.2 to 0.45 c irradiation, there appeared binuclear and polyploid cells, whose number increased with the increase of the dose. Groups of polyploid cells formed the entire polyploid sectors in individual cases. At 0.45 c, extremely interesting mosaic plants were found which had individual portions of rootlet tissues with different chromosome numbers: $2n = 14$, $2n = 28$, $2n = 42$ and $2n = 46$. The sectors with different chromosome numbers were arranged in orderly rows in the rootlets. With a much higher dose of irradiation (1 c) a distinctly damaging effect of irradiation was evident by the segmentation of chromosomes and their retardation in apophases.

In the seedlings from pea seeds irradiated after soaking, there were similar cytological changes beginning with the dose of 0.05 c, a much lower dose of irradiation. The most interesting fact noticed in the experiment with pea variety American Wonder was the transfer of the above cytological changes from irradiation with stimulation doses (0.35–0.40 c) to plants of the next generation, which agrees with the hypothesis put forth later by V.S. Andreev on the inheritable nature of the stimulation effect.

The irradiated pea seeds were sown in Mitcherlich pots in four replications and the yield was determined. More vigorous plant development was observed with irradiation of 0.35 c, and the plants grown from seeds irradiated at 0.05 c were no different from the control plants. The criterion of the estimate was the number of pods (Table 4.50).

Table 4.50. Average number of pods per plant

Dose of irradiation, c	Average number of pods per plant	Dose of irradiation, c	Average number of pods per plant
Control	2.33±0.32	0.45	3.57±0.33
0.05	2.54±0.35	0.55	3.38±0.39
0.10	2.56±0.35	0.65	3.25±0.39
0.20	2.57±0.39	0.75	3.38±0.35
0.25	3.63±0.71	1.00	2.64±0.44
0.35	4.63±0.38		

A biometrical analysis of the results of these experiments confirmed the observations of more pods in the plants which were grown from seeds irradiated in the dose of 0.35 c. The data indicate the effect of x-rays through the entire growth periods of the plants. A.I. Atabekova's research is among the first of that time when Köernike's theory on the presence of stimulation only in the initial developmental stages of plants dominated.

At the K.A. Timiryazev Agricultural Academy [243] the researchers observed the change of growth and development of pea whose seeds were irradiated by the flux of thermal neutrons at 1 and 1×10^{12} neutrons/cm^2. Microsporogenesis took place normally and the pollen germinated well on the artificial medium. The stimulation effect was evident in the phenomenon in that the development of the ovule surpassed the control and the development of the embryonic sac accelerated considerably.

During 1968–1971 in the MSSR, commercial crops were raised from pea seeds which were irradiated with doses of 0.3–1.2 c [152]. At the state farm Zarya Kommunizma of the Kagul'skii region in 1969, pea seeds of the variety Ramonskii 77 irradiated at 1.2 c were sown in three replications on 11.4 ha in plots of 0.95 ha each. The increase in the grain yield was 3.4 q/ha which was 30.9% over the yield from the control. In 1970, at the state farm Dnestr, pea seeds of the variety Belladonna, irradiated at 0.3 c, were sown on 2.75 ha. In this experiment the increase in yield was 5.1 q/ha, i.e., 26.8% over the control.

Bequerel [269] treated pea seeds with uranium nitrate in concentrations of 10^{-4} and 10^{-3} mc/l (the seed germination somewhat reduced in higher concentrations) 24 hours before sowing. Peas treated with uranium nitrate flowered five days earlier than the control. The yield from crops of treated seeds was 10% higher than that of the control.

In France experiments on irradiation of pea seeds were conducted by Vidal. According to his data, the presowing irradiation of pea seeds by ^{60}Co gamma-rays with doses of 0.25–0.5 Krad increased the yield potential of pea by 16–41% with much earlier ripening [362].

In Canada MacKey irradiated air dried pea seeds with 0.3 c and the number of flowering increased by 15%. In 40% of the plants the pods matured considerably earlier; whereas, the total mass of grains obtained from plants grown from irradiated seeds exceeded the control by 24%.

Hungarian experiments conducted with the pea variety Petit Provencal also showed the stimulation effect of the 1.5 c dose, which manifested an almost 20% increase of the yield [288].

Results of experiments on the irradiation of pea seeds are shown in Table 4.51.

Beans. This plant is a favorite object of laboratory radiobiological research. Beans are highly radiosensitive. Evler [279], in his research, noticed the accelerated germination of beans irradiated before sowing. Promcy and

Table 4.51. Consolidated data on the presowing irradiation of pea seeds

Place of research	Year	Researcher	Dose of irradiation	Increase in yield in comparison with the control, %	Change of the biochemical composition	Enhancement of development
All-Union Research Institute of Fertilizers and Agril. Soil Science	1935	L.P. Breslavets, A.I. Atabekova	0.35 c	35	—	Enhancement of ripening
Ural branch of the USSR Academy of Sciences	1958	N.A. Poryadkova	0.30–0.50 c (dry seeds)	—	—	Enhancement of ripening
			0.07–0.08 c (soaked seeds)	28–33	—	-do-
Gorky State University	1961	V.A. Guseva	0.5–1 c	11–37	Increase in the content of phosphorus and nucleic acids	—
	1962		0.5–1 c	3–3		—
MSSR, Collective farm Zarya Kommunizma	1969	V.N. Lysiko, K.I. Sukach, D.M. Goncharenko	1.2 c	30.9	—	—
MSSR, State farm Dnestr	1970	V.N. Lysikov, K.I. Sukach, D.M. Goncharenko	0.3 c	26.8	—	—
France	1959	Vidal	0.25–3 Krad	16–41	—	—
Canada	1970	MacKey	0.3 Krad	24	—	Enhancement of ripening

Drevon [338] and Iven [297] observed the increased and accelerated germination of beans due to low doses of radiation. Köernike [304] and Altman, et al., [262] noticed that, due to effects of low doses of radiation, the growth and development accelerated in plants grown from irradiated seeds. The criteria of enhanced growth and development were the measurements of the root and aerial parts of the plants.

E.I. Preobrazhenskaya [190] subjected the fodder beans variety Chernye russkie to low doses of irradiation; for this variety even a dose like 0.1 c was suppressive.

In this way, generalizing the data on the presowing irradiation of bean seeds, we may note the very high radiosensitivity of this crop, for which, the stimulation effect was observed with irradiation at 0.07 c. Maybe this was one reason that Köernike negated the stimulation effect in his experiments. He worked mostly with beans when it was hardly possible to accurately establish a low dose like 0.07 c, that is, in the initial days of radiobiological research.

Common bean, lentil, lupin and vetch. The insignificant amount of research on the presowing irradiation of these plants of family Leguminaceae, makes it possible to include them in a common group. Work on the irradiation of the common bean dates back to the early development of radiobiology. Promcy and Drevon [338], while irradiating seeds of common bean and lentil, noticed the enhanced germination and development of plants. Rochlin [341] observed more intensive cell division and differentiation of meristemic cells during the irradiation of the common bean.

At the Tomsk State University, researchers have established the stimulation of lentil germination, the increased germination percentage and enhanced development of seedlings from seeds which were subjected to x-ray irradiation [246].

Irradiation of lupin seeds with 1 c [233] enhanced the development and increased by 16% the yield of green mass. We have tried dose variations from 0.25 to 16.0 c, with strength of 720 r/min, during irradiation of lupin seeds by ^{137}Cs gamma-rays. It must be noticed that stimulation for this crop was attributed to a very wide range of doses from 1 to 8 c. The amount of alkaloid contents in the seeds and green mass increased with an increased dose. The increase in the yield of green mass was as high as 27%. However, this one year's data need verification.

Thus, while generalizing data about the presowing irradiation of these pulse crops (Table 4.52), we see that lentil, common bean and vetch have not been studied adequately; the available data are preliminary and confined mainly to information about the germination percentage of seeds, germination energy, enhanced initial developmental stages and so on. Field experiments, in which the yield was taken into consideration, and subsequent large-scale production trials were conducted only on pea, lupin and vetch and were too limited.

Table 4.52. Consolidated data on the presowing irradiation of the seeds of common bean, lentil, lupin and vetch

Place of research	Year, crop	Researcher	Dose of irradiation, c	Increase in yield in comparison with the control, %	Enhancement of development
All-Union Research Institute of Fertilizers and Agril. Soil Science	1934 Lupin	M.M. Tushnyakova, K.A. Vasilevskii	1	16	Enhancement of development
Institute of Plant Physiology of the USSR Academy of Sciences	1959 Lupin	P.A. Vlasyuk	1	20	—
Institute of Biophysics of the USSR Academy of Sciences	1962 Lupin	N.M. Berezina, Riza-Zade	1–2	18–27	—
Kishinev Agricultural Institute	1969 Vetch	K.I. Sukach, V.N. Lysikov	1.2	—	Delay in development

Rye. Rye seeds were first irradiated by Promcy and Drevon [338]. Germinating the irradiated rye seeds at a low temperature (15°C), they obtained contradictory results which showed both enhanced as well as retarded germination of irradiated seeds. However, if the temperature during germination was higher (35–40°C), the irradiation always favorably affected the germination and development of seeds.

L.V. Doroshenko subjected air dried rye seeds to x-ray irradiation for 5, 10 and 20 minutes. The dose given for 20 minutes favorably affected germination from the very first moments [93]. Seedlings from irradiated seeds were complete, whereas only a few seedlings were noticed in the control. In plants raised from irradiated seeds, tillering began two days earlier, stem elongation four days earlier and heading two weeks earlier than in the control plants. Unfortunately L.V. Doroshenko did not consider the yield of irradiated plants. He observed only the paces of the developmental phases of rye due to the effects of irradiation. His results, although from methodologically unsystematic experiments, show the significant enhanced plant development due to the effects of low irradiation doses.

In N.V. Chekhov's experiments [246] at the Tomsk State University, it was shown that the presowing irradiation of air dried rye seeds in a definite dose caused much faster seed germination and more vigorous plant development. Substantial experiments on the irradiation of rye seeds were conducted by L.P. Breslavets, et al. [56, 57]. They irradiated rye seeds which were soaked for 12 hours. Short and long wave x-rays were used for irradiation. Long wave rays worked less sharply, reduced the damages of high doses and increased the stimulation effects of low doses. Harvest and yield estimates were done separately for each plant (Table 4.53).

The number of spikes and number and mass of grains increased almost threefold due to the effects of irradiation at 0.25 c. The stimulation effect of irradiation at 0.5 c was very weak, however, the number of spikes and grains still significantly exceed the control. Doses of 0.75 and 1 c stunted the development of spikes.

Table 4.53. Effect of the presowing irradiation of seeds on the productivity of rye

Dose of irradiation, c	Average number of spikes	Average number of grains			Average mass of grain, g
		normally developed	undersized and diseased	total	
Control	3.7	111.7	35	146.7	5.36
0.25	10.0	255.0	85	340.0	12.52
0.50	7.0	122.0	59	181.0	6.52
0.75	4.5	82.0	37	119.0	4.57
1.00	3.5	80.0	13	93.0	3.18

As a result of the x-ray irradiation of soaked rye seeds, the spikes of some plants surpassed the size of spikes of the control (Fig. 4.3) by almost 1.5 times, but particularly astonishing was the grain size on these spikes, which resembled the grains of polish wheat (*T. polonicum*). In the control, most of the grains had the mass of 15.5 mg, whereas the grain mass was 39.5 mg for spikes from irradiated seeds.

Estimates of the number of spikes and grains per plant showed that the number of spikes increased sharply with irradiation at 0.25 c. As the dose increased to 2 c, it gradually reduced in comparison with the control. Further increases of the dose to 4 and 8 c reduced the number of spikes to three, i.e., it became less than in the control (Table 4.54). Similarly, we must pay attention to the reduced number of diseased grains in the irradiated plants. In the control, of 135 grains, 48, i.e., 35.5% were the diseased; the irradiation of seeds at 0.25 c resulted in the fact that, of 446.5 grains only 63.7, i.e., 14.3%, were diseased. The dose of 0.25 c increases the grain mass threefold. Doses of 0.5 and 0.75 c hold it at a somewhat lower level of stimulation, but the grain mass begins to decrease with 1 c. Examining the data for each plant, the authors found great individual variations of the data within the range of a particular dose. For example, the number of spikes varied from 2 to 24, the number of normal grains from 17 to 930 and the total grain mass from 1.06 to 36.26 mg.

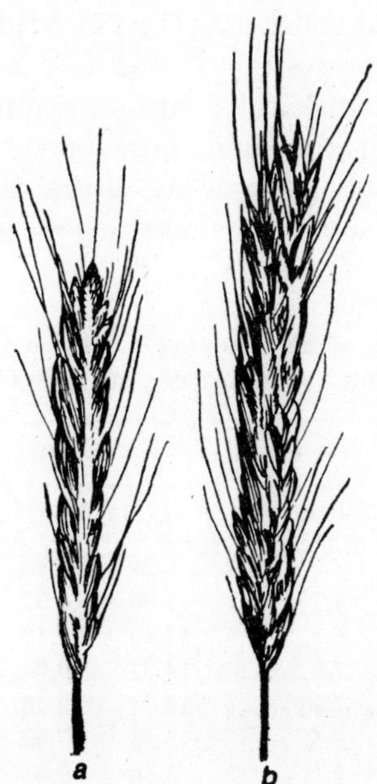

Fig. 4.3. Effect of the presowing irradiation of soaked rye seeds: *a*—control; *b*—irradiated.

Table 4.54. Effects of the presowing irradiation of rye seeds on the number of spikes (per plant)

Dose of irradiation, c	Average number of spikes	Average number of grains			Average grain mass, g
		normally developed	undersized and diseased	total	
Control	4.5	87.0	48.0	135.0	5.29
0.25	11.7	382.8	63.7	446.5	15.96
0.50	8.7	292.5	52.0	334.5	13.43
0.75	8.0	297.0	61.0	358.0	14.08
1.00	7.0	230.5	70.0	300.0	10.74
2.00	4.5	131.0	20.5	151.5	5.58
4.00	3.0	90.0	38.0	328.0*	3.66
8.00	3.0	57.0	59.0	86.0*	3.21

*Correct as per the Original Russian text. An obvious error in the original data—General Editor.

From the data presented above it is evident that x-ray radiation is a stimulator of the growth and development of rye. If we compare all growth, development and productivity curves for the irradiated plants, we may see their complete coincidence. The initial plant growth, diameter of radicles, number of spikes and number and mass of the grains—all these characters appear absolutely identical under the effects of irradiation: with dose of 0.25 c—the curves ascend steeply; with 0.5 and 0.75 c—the curves rise in the same direction, but not that sharply; with 1 c and more the curves fall.

The positive effects of stimulation doses of x-rays are not confined to just increased productivity, the progeny of irradiated plants are influenced as well. These data were obtained by resowing rye seeds taken from irradiated plants (Table 4.55).

Table 4.55. Effects of the presowing irradiation of rye seeds on the development of the second generation of plants

Dose of irradiation, c	Number of spikes		Total grain mass, g		Number of grains	
	1931	1932	1931	1932	1931	1932
Control	4.5	7.3	5.29	7.74	135.0	222.2
0.25	11.7	4.7	15.96	4.93	446.0	168.2
0.50	8.7	6.8	13.43	9.17	343.5	279.2
0.75	8.0	7.8	14.03	10.04	358.0	306.5
1.00	7.0	7.2	10.74	8.78	300.5	277.6
2.00	4.5	5.5	5.58	7.42	151.5	215.7
4.00	3.0	7.1	3.66	9.44	128.0	283.7
8.00	3.0	7.2	3.21	8.18	86.0	306.6

Data in the table show the definite after-effects of x-rays in certain doses as expressed in the increased yield in the second generation. A dose of 0.75 c not only increases the grain mass by 66% in the year of irradiation but also increases the productivity by almost 30% in the next year. Irradiating rye grains with 0.25 c increased the yield by 21% in the year of irradiation but was negative in the second generation. The doses of 4 and 8 c, highly stunted the plant growth in the year of irradiation, but produced a positive effect subsequently.

The possibility of increasing the rye yield by 8–11% through the presowing irradiation of seeds with doses of 0.25–0.50 c was established at the Ural branch of the USSR Academy of Sciences (Table 4.56).

Table 4.56. Yield of rye grown from irradiated seeds (in one plot)

Dose of irradiation, c	Yield, g	In percentage to the control
Control	138.4	100
0.10	143.1	104
0.25	149.3	108
0.50	158.3	111
0.75	133.3	91

Air dried rye seeds of the variety Pionerkaya were irradiated there by ^{60}Co gamma-rays at 1, 2, 3 and 5 c. The author [207] aimed to reveal not only the importance of cumulative doses of irradiation, but also the effect of the dosage strength on the final irradiation effects. Therefore, different strengths of dose were applied, viz., 5, 50 and 500 r/min. Immediately after irradiation the seeds were sown in 1 m^2 plots in five replications (300 grains in each replication). Rye seed irradiation at 1 and 2 c increased the plant productivity. In this case, the lesser the dosage strength, the longer the duration of seed irradiation and the stronger the stimulation effect of these doses. The author concluded that with the same dose of irradiation the stimulation effect of long-term irradiation was stronger than that of the short-term irradiation.

Results of the experiments on the presowing irradiation of rye seeds are presented in Table 4.57.

Wheat. The first experiments on the irradiation of air dried wheat seeds were conducted by Schwartz, Czepa and Schindler, who could not detect any stimulation effect of irradiation on neither germination nor on the development of seedlings [349].

According to the data from G. Frolov [237], the presowing irradiation of wheat seeds increased the yield by 60% in comparison with the control. B.I. Tsuryupa [242] irradiated durum wheat seeds and noticed the increase

Table 4.57. Consolidated data on the presowing irradiation of rye seeds

Place of research	Year	Researcher	Dose of irradiation	Increase of yield in comparison with the control, %	Enhancement of development
All-Union Institute of Plant Industry	1929	L.V. Doroshenko	20 min.	—	Enhancement of germination, tillering and heading
All-Union Research Institute of Fertilizers and Agricultural Soil Science	1931	L.P. Breslavets, A.S. Afanas'ev, G.B. Medvedeva	0.25 c	14.3	—
Tomsk State University	1932	V.P. Chekhov	1 c	—	Enhancement of seed germination, more vigorous development of seedlings
Ural branch of the USSR Academy of Sciences	1962	V.N. Savin	1 c (strength of the dose 5 r/min)	12	—
Ural branch of the USSR Academy of Sciences	1963	E.I. Preobrazhenskaya	0.5–1 c	4–11	—

in the energy of seed germination. According to his data, the reproductive organs gave more increase than the vegetative organs. In individual cases in his experiments, the increase of wheat grain mass was as high as 60%.

At the Leningrad Agricultural Institute [102] air dried wheat seeds of the variety Diamant were irradiated with doses of 0.25, 0.5, 1 and 1.5 c, with the strength of 250 r/min. Field experiments were conducted in five replications on small plots, each with an area of 0.72 m². The optimal dose of irradiation was 0.5 c and the grain yield increased by 12.7–23% with this dose.

Research workers of the Gorky University studied the stimulation effect of low dose of irradiation on wheat, through chronic irradiation of the seedlings. The experiments showed that, with the strength of dose below 1 r/day, the stimulation effect was observed to manifest in the increase of grain yield by 27.8–30.6%, when the plants were grown on well fertilized land. Stimulation of plants growing on unfertilized land, was manifest considerably poor even with lower irradiation dose.

At the All-Union Research Institute of Fertilizers and Agricultural Soil Science, air dried wheat seeds were irradiated with 0.5, 1, 2 and 3.5 c. With 3.5 c the yield increased by 10%. However, this increase was not observed when the experiment was repeated [131].

While studying the after-effects of presowing irradiation, the authors revealed the possibility of getting considerably more increased yields in the second generation than in the year of irradiation (Table 4.58).

The after-effects of irradiation reflected in the increased grain yield in the second generation was because of the increase in the number of spikes and not their size (Table 4.59).

It must be mentioned that neither of the doses under study produced any effect on the increased germination percentage of wheat seeds or the energy of their germination.

Table 4.58. Effects of the presowing irradiation of wheat seeds on the yield of second generation

Dose of irradiation, c	1957		1958	
	g/pot	%	g/pot	%
Control	4.4±0.2	100	17.6±1.0	100
0.5	4.6±0.3	104	18.6±0.2	105
1.0	4.6±0.2	104	20.6±0.2	117
2.0	4 7±0.1	106	21.6±0.2	123
3.5	4.9±0.1	110	20.6±0.7	117

Table 4.59. Effects of the presowing irradiation of wheat seeds on absolute mass of grains and number and average mass of spikes of the second generation

Dose of irradiation, c	Absolute mass of grains per pot, g	Number of spikes per pot	Mass of one spike, g
Control	32.4	24	1.4
1.0	—	27	1,4
2.0	33.0	28	1.4
3.5	32.0	—	—

In the Ural branch of the USSR Academy of Sciences, wheat seeds of the variety Lutescence 62 were subjected to presowing irradiation by ^{60}C gamma-rays at 1, 2, 3 and 5 c and each dose was applied in three strengths—5, 50 and 500 r/min. The irradiated seeds were sown in five replications of 300 seeds each in 1 m^2 plots. The irradiation of seeds with 2 c increased the grain yield by 8–14%, depending on the strength of the dose; the lower the strength of the dose, the longer the duration of irradiation and the stronger the stimulation effect of irradiation. With the irradiation of wheat seeds at 5 c, the damaging effect intensified with the reduction of dosage strength.

Süs [357] irradiated air dried seeds of the spring wheat variety Lichti with 0–0.8 c at the strength of 20 r/min and observed the stimulation effect. The result was the grain yield increased by 3–5% in comparison with the control.

Seeds of the soft wheat variety Gori were irradiated at the Oak Ridge Laboratory [USA]. The stimulation effect was observed through irradiations at 0.5, 1 and 2 c.

Kersten [308] subjected wheat seeds to x-rays at 18 KV and used a 0.035 mm thick aluminum filter in a 0.04 mm thick polythene sheet. The object of the research was to study the biochemical changes in the plants caused by irradiation. Material for the analysis was prepared in the following way. Seeds were soaked for 12 hours in distilled water so their moisture content was raised to 28%. The soaked seeds were irradiated and germinated in darkness over seven days, then they were dried at 48°C. Diastase activity, total content of sugars and dry mass were determined from dried seedlings. The author examined all changes vis-a-vis the irradiation time. Diastase activity was determined on the basis of the amount of soluble starch per 1 g of dry mass of the seedlings.

Irradiating soaked seeds for five seconds stimulated the amylase activity and sugar content, while the intensity of respiration and water content in the seeds were below the control for all periods of irradiation.

Vidal [362] irradiated wheat seeds of the variety Wilmoren 27 by ^{60}Co gamma-rays and, without mentioning the conditions of irradiation, he presents data showing up to 28% increase in the yield of wheat grown from irradiated seeds. Jonson studied the effects of ionizing irradiation on dry and soaked wheat seeds [299].

K.V. Yanchenko [259] irradiated air dried wheat seeds and observed the variations of effects of the doses depending on the interval of time between irradiation and sowing. He irradiated seeds of the variety Kamalinka, which were later stored for 18 days before sown in the soil. As a result, the wheat yield increased by 31%, whereas, when freshly irradiated seeds in the same dose were sown, there was no increase in yield. Analogous results were observed for the variety Diamant, whose irradiated seeds were stored for 25 days before sowing.

K.V. Yanchenko's observations show that the interval of time between irradiating and sowing wheat seeds plays a decisive role in the stimulation of effect and, therefore, deserves greater attention and further all-round study. This is particularly important because before there were no precisely established doses of irradiation from which we could get quite stable results for wheat, which is the most valuable of agricultural crops.

At the Georgian Research Institute of Agriculture F.A. Dedul' [84] conducted two years of experiments on the presowing irradiation of seeds of the winter wheat variety Bezostaya 1. The experiments were conducted to determine the optimal storage periods of irradiated seeds before sowing.

Air dried seeds with 12.8 and 12.9% moisture content were irradiated by x-rays in doses of 1, 5, 10 and 20 c, with a strength of 28 r/min using the RUM-II apparatus. They were also irradiated by ^{60}Co gamma-rays in the same doses at the strength of 380 r/min, after which the seeds irradiated with both x-rays and gamma-rays were sown on three different days—on the 2nd, 30th and 360th day. Data on the wheat yields of seeds sown on these dates are presented in Table 4.60.

An increased wheat yield was obtained because of better tillering, increased mass of 1,000 grains and much the higher test weight of the grains. Results of the experiment showed that storing irradiated seeds for one year reduces the stimulation effect.

Since 1971, extensive large-scale commercial tests on the presowing irradiation of seeds irradiated on the Kolos unit were launched in Pavlodar province, where 312 tons of seeds, including wheat, were irradiated. The total area sown with irradiated seeds in the province was 7,000 ha, of which 409 ha were sown with wheat seeds. The increased wheat yield as a result of irradiation with 0.5 c was 10.3%. In 1972 in the same province, extensive large-scale experiments under production conditions were conducted to establish the optimal storage periods for irradiated seeds to increase the yield [157, 158].

Table 4.60. Grain yield of the winter wheat variety Bezostaya 1 seeds sown on the 2nd, 30th and 360th day after gamma-irradiation

Type and dose of irradiation, c	Yield of grain with 14% moisture content, q/ha	Deviation from the control	
		q/ha	%
	Sown on 2nd day after irradiation		
Control	44.9	—	—
^{60}Co gamma-rays:			
1	41.2	− 3.7	− 8.0
5	39.7	− 5.2	−12.0
	Sown on 30th day after irradiation		
Control	49.0	—	—
^{60}Co gamma-rays:			
1	65.4	+16.4	+33.5
5	53.4	+ 4.8	+ 9.8
	Sown on 360th day after irradiation		
Control	55.1	—	—
x-rays:			
1	51.8	− 3.3	− 6.0
5	48.2	− 6.9	−12.5
^{60}Co gamma-rays:			
1	51.0	− 4.1	− 7.4
5	53.0	− 2.1	− 3.8

I.Ya. Malishchuk irradiated seeds of the winter wheat variety Belotserkovskaya 198 by long and short wave x-rays and tried a very wide range of doses—from 0.125 to 4 c. In his experiments the optimal doses were 3 and 4 c, which had a slight retarding effect on the germination energy and increased the seedling mass during germination. With 0.75, 2 and 3 c of long wave x-rays, from four years' data the average increase of grain yield was to the tune of 3.8, 5.12 and 4.20 q/ha or 11.8, 17 and 13.9% respectively. When seeds were irradiated with short wave x-rays at 0.75, 2 and 3 c, there was increase in grain yield by 21.2, 29.4 and 26.9% respectively.

According to MacKey's data presented at the Geneva conference on the peaceful utilization of atomic energy (in 1971), there was 15% increase of wheat grain yield through the irradiation of seeds with 1 c in Canada from 1968 to 1970.

In Hungary [336] much higher indices in comparison with the control were obtained from field experiments on the irradiation of air dried seeds of the wheat variety Bezostaya 1, with 2 c: number of leaves—5.13 (control—4.91); average width of leaves—1.12 cm (1.02 cm); average weight of green mass obtained from 1 m^2—405.76 g (291.95 g); average weight of dry mass—62.61 g (55.56 g).

While summing up the results of experiments, it must be mentioned that the work on the presowing irradiation of wheat seeds has great national economic importance and must be extended to their maximum to study a much wider assortment as well as to determine the optimal doses of irradiation in the context of different zones of cultivation and storage periods of the irradiated seeds before sowing.

Results of experiments on the irradiation of wheat seeds are given in Table 4.61.

Barley. In the initial works of Schwartz, Czepa and Schindler [349] and Patten and Wigoder [335], who irradiated air dried barley seeds of the variety Winner at the dose of 8 NED, no stimulation was observed. N.V. Chekhov [246] found an increased germination percentage and enhanced development of seedlings as a result of the irradiation of air dried and sprouted seeds.

O.K. Kedrov-Zikhman and N.I. Borisova [131] irradiated air dried barley seeds of the variety Winner with doses of 0.1, 0.2, 0.3, 0.4, 0.5, 0.6, 0.7, 1, 2 and 3.5 c. They studied the effects of irradiation over a period of three years; the irradiated seeds were sown in pots. A small, but significant stimulation irradiation effect appeared on the barley seeds almost uniformly during all three years (Table 4.62).

As may be seen from the data presented in the table, the presowing irradiation of barley seeds at 0.1–0.5 c increased the grain yield by 7–12% during the three years and the optimal dose varied from year to year depending on environmental conditions, but it did not exceed 0.5 c. For example, in 1957, the increased yield was obtained through irradiation with 0.5 c, in 1958—0.3–0.4 c and in 1959—0.1–0.2 c. The estimate of the vegetative mass of barley showed different irradiation effects, which appeared in a much wider range of doses.

A study on the effects of presowing irradiation on second generation seeds showed the absence of the stimulation effect in the second generation. The authors also analyzed the stimulation effects of x-ray doses during repeated irradiation, for which they used the same doses as in the first irradiation. There was no yield increase through repeated irradiation of barley seeds.

The dose of 0.5 c was found to be optimal for the presowing irradiation of the barley variety Winner in the experiments of N.G. Zhezhel' for Leningrad province (Table 4.63) [102]. Among the biochemical changes which have occurred in the grain as a result of the presowing irradiation of barley seeds, the increased protein content has great practical value. In the control the protein content was 11.1% and 12.0% in the irradiated grain.

In 1958, experiments on the presowing irradiation of barley were started in the Kirghiz SSR. The effects of different doses of irradiation on the survival rate, productivity and protein content in the grain were studied in

Table 4.61. Consolidated data on the presowing irradiation of wheat seeds

Place of research	Year	Researcher	Dose of irradiation	Increase of yield in comparison with the control, %	Change of biochemical composition
Azovo-Chernomorskii Agricultural Institute	1934	B.P. Tsuryupa	0.5–1.0 c	60	—
Oak Ridge Laboratory (USA)	1934	Kersten	0.51; 2.5 c	Stimulation	Activation of amylase; increase in sugar content by 8–14%
K.A. Timiryazev Agricultural Academy	1936	G. Frolov	1.0–2.5 c	60	—
Gorky State University	1955–1956	V.A. Guseva	Chronic irradiation on gamma-field	27.8–30.6	—
K.A. Timiryazev Agricultural Academy	1957	O.K. Kedrov-Zikhman, N.I. Borisova	2 c	23	—
Krasnodar Pedagogical Institute	1959	K.V. Yanchenko	1.5 c (with subsequent storage during a period of eight days)	31	—
France	1959	Vidal	1 Krad	28	—
GDR	1959	Süs	0.8 Krad	3.51	—

Kishinev Agricultural Institute	1968–1971	V.N. Lysikov, K.I. Sukach, D.M. Goncharenko	1.0–1.5 c	15-18	—
Ukrainian SSR	1972	I.Ya. Malishchuk	Long wave x-rays	11–17 21–29	— —
Canada	1968–1971	MacKey	1 Krad	15	—
Hungary	1968	Pannonhalmi	—	11.5	—
Pavlodar Provincial Directorate of Agriculture	1970–1971	Yu.A. Martem'yanov, A.V. Kalashnikov	1 Krad	10.3	—

Table 4.62. Grain yield of barley grown from irradiated seeds

Dose of irradiation, c	1957		1958		1959	
	g/pot	%	g/pot	%	g/pot	%
Control	22.0±1.0	100	20.1±0.1	100	18.2±0.6	100
0.1	—	—	—	—	20.4±0.1	112*
0.2	—	—	—	—	20.4±0.3	112*
0.3	—	—	21.9±0.2	109*	19.3±0.6	106
0.4	—	—	21.5±0.9	107**	—	—
0.5	24.3±0.6	110*	21.0±0.4	105	18.2±0.5	100
0.6	20.3±1.0	92	20.3±1.0	101	—	—
0.7	—	—	20.1±1.2	100	14.5±1.1	80
1.0	22.5±2.6	102	—	—	—	—
2.0	20.1±2.3	94	—	—	—	—
3.0	22.0±2.0	103	—	—	—	—

*Difference significant at 95%.
**Difference significant at 90%.

Table 4.63. Effectiveness of the presowing irradiation of barley seeds by ^{60}Co gamma-rays

Dose of irradiation, c	Yield per plot, kg	Yield in comparison with the control, %
Control	2.46	100.0
0.25	—	—
0.5	2.66	108.0
1.0	—	—
1.5	2.56	103.2

these experiments. It was established that barley is more radiosensitive than wheat under Kirghiz SSR conditions. The spring barley variety Nutans 45, when irradiated with 1 c, gave a 13–15% increase of grain yield. The nitrogen and phosphorus contents of the grains increased with an increased dose.

In 1971, large-scale tests on the presowing irradiation of barley were conducted with two varieties, viz., Kambainer and Nutans 45, whose seeds were irradiated with 1–3 c (Table 4.64).

As a result of research conducted at the Research Institute of Agriculture of Kirghiz SSR [183], it has been established that a stimulation effect was produced by doses of 0.5, 1 and 2 c: increased grain yield of the variety Nutans 45 in small plot experiments was to the tune of 7–15% in comparison with the control. In these doses, the nitrogen and phosphorus contents in the grains of the variety Nutans 45 increased by 17.8–69.9% and 43–85.7% respectively.

A large-scale trial was also conducted in the Kazakh SSR on the presowing irradiation of the barley variety Omskii 13709 on an area of 200 ha (100 ha—experimental, 100 ha—control) in four replications. Seeds were irradiated at 0.5 c. As a result of irradiation there was an increase in the grain yield of 2 q/ha (the control yield was 9 q/ha), an increase of 22.2% over the control. Mathematical analysis of the experiment's results showed a significant yield increase.

The German radiobiologist Süs, during 1959 and 1960, investigated the effects of presowing gamma-irradiation on seeds of the spring barley variety Donariya with a moisture content of 13.2%. He irradiated seeds in doses of 0.04–16 c, with the strength of 20 r/hour. Seeds were sown seven and 14 days after irradiation. The results of two years of his experimental data are presented in Table 4.65.

Results of much later research by Süs (1964–1967) on the irradiation of air dried seeds of the spring barley varieties Early and Union are given in Table 1.1.

The data presented by Süs show the great variability in the increases of

Table 4.64. Results of large-scale tests on the presowing irradiation of barley seeds

Place where experiments were conducted	Variety	Dose of irradiation, c	Plot size, ha	Replications	Yield, q/ha	Increase of yield in comparison with the control, %	
						q/ha	%
Collective farm Tendik of Kochkorskii region	Kombainer	Control	1	3	21.3	—	—
		1	1	3	22.5	+1.2	105.6
		3	1	3	23.1	+1.8	108.9
State farm Kommunizm of Kochkorskii region	Nutans 45	Control	2.7	2	21.5	—	—
		1	2.7	2	21.5	—	100.0
		3	2.7	2	22.3	+0.8	103.7

Table 4.65. Effects of the gamma-irradiation of spring barley seeds on yield, % in comparison with the control

Storage period of irradiated seeds, days	Dose of irradiation, Krad				
	0.04	0.08	0.8	8	16
			1959		
7	108.2	159.6	154.7	123.8	121.3
14	110.8	158.5	154.0	130.8	138.7
			1960		
7	105.8	116.9	105.8	106.4	107.0
14	106.2	115.3	117.3	112.4	109.3

spring barley yields from year to year, depending on variety. In Süs' research it is also interesting that in all experiments he used very low doses with very low strength.

According to reports from MacKey, the 1969 Canadian experiments on the irradiation of barley seeds by ^{60}Co gamma-rays showed very little yield increases.

Experiments on the chronic irradiation of spring barley conducted by Glubrecht revealed a stimulation which was distinctly evident in the increase of grain and straw (Tables 4.66 and 4.67). The data are very interesting and show the increased grain and straw yields through the irradiation of seeds in different doses. In Glubrecht's opinion, we get much higher and more stable yield increases with chronic irradiation of plants than through acute irradiation.

It is evident from the table that more increased grain yields were observed with much lower dose of irradiation than increased straw yields; the optimal stimulation dose for grain was 0.122–0.94 c, while the optimal dose for straw was 0.98–1.97 c, and a still higher dose (to 3.95 c) was required to increase the yield of chaff.

In 1970, the international agency, FAO/MAGATE conducted an experiment in Austria on the presowing irradiation of spring barley seeds on an international scale and following uniform methodology. Two barley varieties were selected for the experiment: one was the extensively cultivated variety Winner and the second was a local variety cultivated where the experiment was conducted. Quite detailed and well planned instructions envisaging completely uniform and timely experiments were sent to all participants in the international field experiment with a random block design. The research was undertaken to reveal the stability of the yield increases through the presowing irradiation of seeds and to solve the problem of the possibility and profitability of applying the presowing irradiation of seeds in practical

Table 4.66. Grain yield of the spring barley variety Brown Visa through chronic irradiation of seeds

Dose of irradiation, Krad	No. of spikes per 1 m²	Grain yield (dry mass)				1,000 grain mass		No. of grains in the spike	
		g/m²	%	mg/spike	%	g/m²	%	number	%
Control	399	299	100	561	100	31.77	100	20.06	100
15.8	240	4	2	15	3	22.20	70	0.29	5
7.9–15.8	357	12	5	34	6	30.71	97	1.00	5
3.95–7.9	532	24	11	46	8	32.08	101	1.25	6
1.975–3.95	531	109	48	207	37	34.47	109	6.60	33
0.988–1.975	487	177	78	365	65	35.15	110	10.70	53
0.494–0.988	**586**	245	103	471	84	**36.22**	115	14.40	72
0.243–0.495	442	259	113	**585**	104	34.88	110	18.10	90
0.122–0.243	509	**302**	**132**	**594**	**105**	33.49	106	**18.70**	**92**

Table 4.67. Straw yield of the spring barley variety Brown Visa through chronic irradiation of plants

Dose of irradiation, Krad	Straw		Chaff		Total		Straw		Chaff		Total	
	g/m²	%	g/m²	%	g/m²	%	mg/spike	%	mg/spike	%	mg/spike	%
Control	209	100	52	100	261	100	524	100	130	100	654	100
15.8	88	42	36	69	124	48	366	70	150	115	516	80
7.9–15.8	146	70	44	85	190	73	410	78	123	95	533	82
3.95–7.9	205	98	73	140	278	106	386	74	137	105	523	80
1.975–3.95	207	99	97	187	**304**	**117**	390	74	183	141	573	88
0.988–1.975	**226**	**108**	70	135	296	113	**474**	**91**	144	111	**608**	**95**
0.494–0.988	192	92	70	135	262	100	330	63	120	92	450	69
0.243–0.494	144	69	73	140	217	83	326	62	**165**	**127**	491	75
0.122–0.243	165	79	**82**	**158**	247	95	334	64	161	124	495	76

agriculture. To evaluate the results and their processing, a commission of experts was constituted and they were invited from different countries to participate in the experiments.

However, during the past two years, in spite of the very well planned methodology, there was no significant yield increase. From our point of view, the object selected for this international experiment was most inappropriate, i.e., barley. This plant, as may be seen from the consolidated data presented by us (Table 4.68), firstly has not been studied adequately and secondly it gives extremely variable results in field experiments and requires, like wheat, exhaustive investigations to reveal the specific modifying factors which cause the unstable results.

Corn. This is one of the most valuable fodder crops. The fodder values of corn grown for silage increase depending on the degree of maturity in the cobs. The maximum quantity of fodder units from one hectare was obtained in the milky-dough maturity stage of plants. However, in the central provinces of the USSR, corn does not always achieve this maturity stage because of early fall frosts.

With the experimental data that the presowing irradiation of seeds almost always results in accelerated plant growth and development, several researchers, independently of each other and in different regions of the country, began to experiment on the presowing irradiation of corn seeds with the purpose of increasing the green mass yield and accelerate the growth and development of plants grown from the irradiated seeds. In the southern regions, similar experiments were conducted to determine the possibility of increasing grain yield by this method.

The experiments on the presowing irradiation of corn seeds of the varieties Voronezhskaya 76, Sterling and hybrid Bukovinskii 3 were conducted by us in collaboration with other researchers at the Institute of Biophysics of the USSR Academy of Sciences [22, 24, 25, 27, 29, 35, 163]. The work was started with a test of the effects of a wide range of doses—from 0.1 to 40 c. The last dose was found to be lethal, killing the corn seedlings on the 13th day of their development. Under laboratory conditions the range of stimulation doses varied from 0.2 to 1 c. However, under field conditions this range significantly changed and doses of 0.5–2 c were found to be the stimulation doses and for large-scale production conditions, doses in a much narrower range, viz., 0.5–1 c were recommended as optimal.

In three years of field experiments in Moscow province, the most stable and optimal increase of yield—from 8 to 28% in comparison with the control—and acceleration of corn development were obtained through irradiating seeds with 0.5 c. Positive results from these experiments served as a basis to conduct large-scale trials on the presowing irradiation of corn under commercial farming conditions. At the initiative of the State Committee on the use of atomic energy of the USSR, this trial was conducted from 1961 to

Table 4.68. Consolidated data on the presowing irradiation of barley seeds

Place of research	Year	Researcher	Dose of irradiation	Increase of yield in comparison with the control, %	Change of biochemical composition
All-Union Research Institute of Fertilizers and Agril. Soil Science	1957–59	O.K. Kedrov-Zikhman	0.3–0.5 c	7–10	—
Leningrad Agriculturul Institute	1963	N.G. Zhezhel'	0.5 c	3–8	Increase of protein content in the grain by 9%
Research Institute of Agriculture of the AzSSR	1970	R.R. Riza-Zade	0.3 c	15	—
Kirghiz Institute of Agriculture	1958–1971	A.S. Sultanbaev, L.A. Sergeeva	1–3 c	3–8	Much higher phosphorus content in the grain
Provincial Directorate of Agriculture of Pavlodar Province	1971	A.V. Kalashnikov, Yu.A. Martem'yanov	0.5 c	22.2	—
GDR	1959–1960	Süs	0.08–0.8 Krad 0.08–0.8 Krad	54–59 15–17	—
GDR	1968	Glubrecht	Chronic	32	—

1963 by the Institute of Biophysics of the USSR Academy of Sciences in the Zaokskii State farm of Moscow province.

First class seeds of the corn variety Sterling from the harvests of 1960, 1961 and 1962, with initial moisture content of 10%, were taken for irradiation. Irradiation was done with 0.5 c, at the strength of 700 r/min. on the mobile gamma-unit GUPOS charged with ^{137}Cs seven days before sowing directly in the state farm. Uniform irradiation was maintained with an error level of up to 15%. In 1961, a total of 1,250 kg of seeds were irradiated and sown in a 60 ha area with highly fertile soils in the flood plain of the River Oka. Measurements on the height, periods of flowering, cob formation, number of cobs and other indices were duly recorded for mature plants. The yield components were analyzed before harvesting and the weight ratio of the individual organs of the plants was determined in the total green mass yield. Besides, the changes in the biochemical composition and fodder values of individual organs of plants due to irradiation were also determined.

At the beginning of vegetative growth, the irradiated plants were 7–10 cm taller than the controls. However, this difference narrowed toward the moment of mass flowering and by harvest the plant height was equal (Fig. 4.4). The development of irradiated plants, as is evident from the data presented in Table 4.69, was faster than that of the control. The number of flowered plants through irradiation with ^{137}Cs gamma-rays on August 9 comprised 10.6%, whereas it was only 2.5% in the control plants. Among the irradiated plants 37.6% plants developed full tassels and 25% in the control (Fig. 4.5). The development of cobs in the irradiated plants was also faster than in control, and the number of large cobs in the irradiated plants was

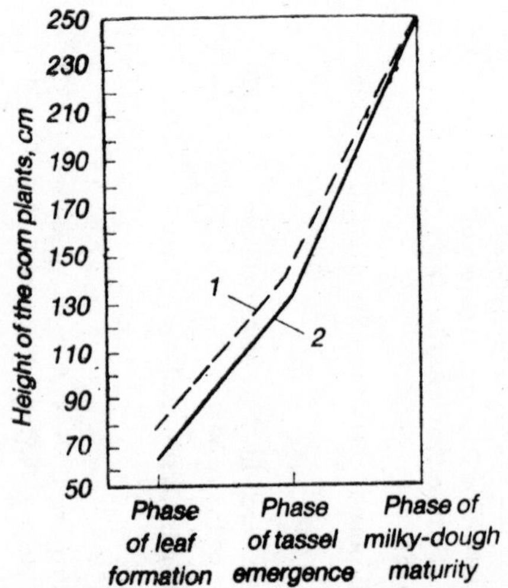

Fig. 4.4. Height of a corn plant irradiated with 0.5 c (*1*) and that of the control plants (*2*).

Table 4.69. Flowering phases of the irradiated plants

Dose of irradiation, c	No. of plants examined	Without tassels		Tassel emergence		Tassels developed		Flowering	
		number	%	number	%	number	%	number	%
Sterling, 1961									
Control	200	60	30	85	42.5	50	25	5	2.5
0.5	292	31	10.6	120	41.1	110	37.6	31	10.6
Sterling, 1962									
Control	200	50	25	54	27	53	26.5	43	21.5
0.5	200	17	8.5	53	26.5	67	33.5	63	31.5
Sterling, 1963									
Control	200	93	46.5	52	26.0	24	12.0	31	15.5
0.5	200	92	46.0	38	19.0	29	14.5	41	20.5
Hybrid Bukovinskii 3, 1963									
Control	200	44	22	49	25	—	—	107	53
0.5	200	22	11	44	22	—	—	134	67

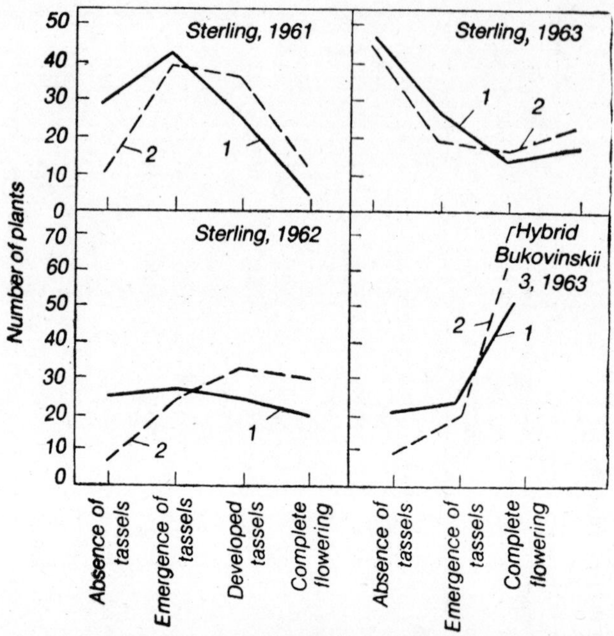

Fig. 4.5. Flowering phases of corn through the presowing irradiation of seeds: *1*—control; *2*—irradiation.

considerably higher than in the control. About 29% of the irradiated plants had very large cobs weighing more than 400 g, whereas only 17% of the control plants had cobs with such mass. The plants grown from irradiated seeds were distinguished by their ability to produce several cobs on one plant and this tendency was observed in the experiments of previous years. On August 9 the number of cobs on different plants was counted (Table 4.70 and Fig. 4.6). The number of multicob plants was considerably higher among the irradiation than in the control: 9% of plants grown from irradiated seeds had four cobs per plant, 30% had three cobs, while in the control such plants were only 0.5 and 5.5% respectively. This increase in the number of cobs through presowing irradiation of corn seeds was the result of awakening of additional generative buds. Quite often numerous underdeveloped cobs form emerging "bunches" instead of a cob from the leaf axis (Fig. 4.7).

Yield components were analyzed before harvesting. In an average sample of 200 plants the total mass of each plant and of its constituent were determined. Thus, we worked out specific value of each constituent in the formation of total yield. The major constituent part of the yield of green mass at harvest comprised stems, cobs and cob-scales or sheaths. The mass of cobs and scales grows significantly with irradiation. The green mass yield of the

Table 4.70. Estimate of corn plants with different number of cobs

Dose of irradiation, c	No. of plants examined	Without cobs		With one cob		With two cobs		With three cobs		With four cobs		With two large cobs	
		No.	%	No.	%	No.	%	No.	%	No.	%	No.	%
					1961								
Control	200	30	15	90	45	68	34	11	5.5	1	0.5	0	0
0.5	221	6	3	25	13	92	45	60	30	18	9	20	10
					1962								
Control	200	130	65	25	12.5	40	20	—	—	5	2.5	—	—
0.5	200	93	46.5	43	21.5	58	29	—	—	6	3.0	—	—

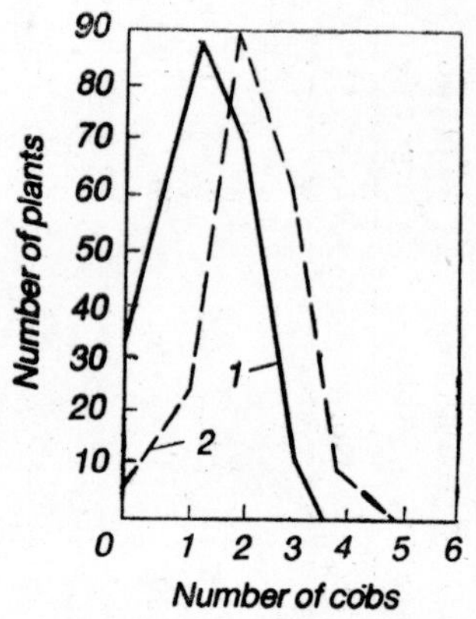

Fig. 4.6. Estimate of corn plants with a different number of cobs: *1*—control; *2*—irradiation.

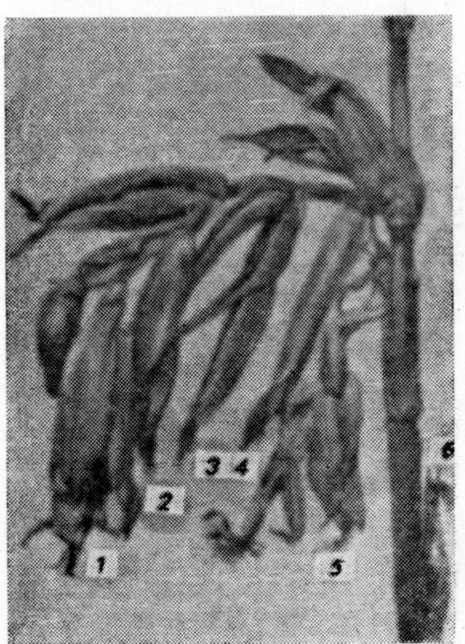

Fig. 4.7. Multicob habit in corn as a result of the presowing irradiation of seeds (six cobs).

control was 321 q/ha, whereas the green mass yield from plants grown from irradiated seeds was 389 q/ha. In this way, the presowing irradiation of seeds with 0.5 c made it possible in 1961 to get an additional 68 quintals of green mass per hectare, which amounted to 21.2% of the control (Table 4.71). Apart from the study on the weight ratio of the green mass [184], a biochemical analysis was also done to determine the chemical composition of individual corn plant organs (Table 4.72). Analysis of data show the content of

Table 4.71. Yield of corn and ratio of organs in it with presowing irradiation of seeds by ^{137}Cs gamma-rays

Dose of irradiation, c	Yield			Stems			Leaves			Cobs			Cob-sheaths		
	q/ha	In percentage of control	Significance of difference	q/ha	In percentage of control	Significance of difference	q/ha	In percentage of control	Significance of difference	q/ha	In percentage of control	Significance of difference	q/ha	In percentage of control	Significance of difference
Sterling, 1961															
Control	321±10	100	—	123	100	38	45	100	14	59	100	18	94	100	29
0.5	389±16	121.2	5.4	137	111.3	35	40	88.9	10	87	147.4	23	125	132	32
Sterling, 1962															
Control	115.2±4.2	100	—	65.1	100	56.5	24.8	100	21.5	25.3	100	22	—	—	—
0.5	134.6±4.8	117.0	3.1	74.0	114	55.0	27.0	109	20.0	33.6	133	25	—	—	—
Sterling, 1963															
Control	478±18	100.0	—	340	100	71.1	103.7	100	21.7	34.4	100	7.2	—	—	—
0.5	551±15	115.3	3.1	379	111.5	68.8	124.5	120.2	22.6	47.4	138	8.6	—	—	—
Hybrid Bukovinskii 3, 1963															
Control	366±19	100	—	196	100	53.6	85	100	23.2	85	100	23.2	—	—	—
0.5	449±20	122.7	3.0	25.2	128	56.0	85	100	19.0	112	132	25.0	—	—	—

Table 4.72. Biochemical characteristics of the yield of corn plants grown from irradiated and unirradiated seeds (1961)

Treatment	Yield, q/ha	Content, % to raw mass						Quantity, kg per hectare						Quantity, %					
		Proteins	Fats	Cellulose	Ash	NES	Sugar	Proteins	Fats	Cellulose	Ash	NES	Sugar	Proteins	Fats	Cellulose	Ash	NES	Sugar
Control:																			
stems	123	1.27	0.33	5.28	1.24	9.80	4.40	156.2	40.6	649.4	152.5	1205	541.2	100	100	100	100	100	100
leaves	45	3.73	1.00	5.32	6.06	11.62	0.79	167.8	45.0	241.2	272.2	523	35.5	100	100	100	100	100	100
cobs	59	2.19	0.52	2.88	0.59	11.82	4.61	129.2	30.7	169.9	34.8	697	271.9	100	100	100	100	100	100
cob-sheaths	94	1.40	0.33	2.98	1.40	8.12	4.51	131.6	31.0	280.1	131.6	763	423.9	100	100	100	100	100	100
total	321	8.59	2.18	16.46	9.29	41.36	14.31	584.8	147.3	340.6	591.1	3188	1272.5						
Irradiation:																			
stems	137	1.58	0.41	0.55	0.96	8.98	5.06	216.4	56.2	760.3	131.5	1230	698	138.5	138.5	117.1	86.2	102.0	128.1
leaves	40	4.56	1.05	5.35	3.82	10.58	0.91	182.4	42.0	214.4	152.8	423	36	108.7	93.3	88.8	56.0	80.8	102.5
cobs	87	2.87	0.69	3.31	0.74	14.88	4.98	249.7	60.0	288.0	64.4	1294	438	193.5	195.4	169.5	185.1	185.6	159.4
cob-sheaths	125	1.27	0.41	2.97	0.99	7.49	4.51	158.7	51.2	371.2	123.7	936	563	120.6	165.2	132.5	94.0	122.6	133.0
total	389	10.28	2.56	12.18	6.51	41.93	15.46	807.2	209.4	1633.9	472.4	3883	1736	—	—	—	—	—	—
Average % of control														138	142	122	88	122	135

proteins, fats, nitrogen-free extracts including sugars and cellulose in different organs and the changes in the content of these substances with irradiation. Calculations of the various indices of yield and biochemical composition in terms of fodder units are shown in Table 4.73. In this way, the green mass of corn grown from irradiated seeds contains 7,209 fodder units, while the green mass of the control plants contains only 5,765 fodder units, i.e., the presowing irradiation of corn seeds with 0.5 c provided an additional 1,444 fodder units of green mass per hectare. The protein content in one fodder unit of irradiated green mass was 70.5 g, but 64.7 g in the control, which also increases the fodder value of the green mass grown from irradiated seeds.

Large-scale commercial crops of the corn variety Sterling were repeated in 1962 and 1963 and phenological observations, yield estimates and evaluations of the fodder value of the green mass of corn were done as per the same methodology that was followed in 1961 [86].

Besides, the effect of the presowing irradiation of seeds was tested on the hybrid Bukovinskii 3, which was released for cultivation in the Moscow province. The data from 1962 and 1963 mainly repeated the phenomena that were observed with the presowing irradiation of the variety Sterling in 1961. As evident from Table 4.70, the total yearly yield of the green mass of corn was 121.2, 117.2, 115.3 and 120.7% in comparison with the control. Good reproducibility of the stimulation effect with the presowing irradiation of corn seeds made it possible to recommend this measure for tests on progressive, advanced farms of the Moscow province.

Air dried corn seeds of the varieties Sterling, Voronezhskaya 76 and VIR 25 were irradiated at the Institute of Biology of the Academy of Sciences of Latvian SSR [132–135]. The initial moisture content of the seeds was about 14%. The trials included the following doses of irradiation: 0.5, 2, 4, 6 and 8 c. The effects of irradiation were judged on the basis of germination and the dynamics of plant growth and development under laboratory conditions. Phenological observations and yield estimates were done in field experiments. Besides, the course of individual physiological and biochemical processes was keenly observed in the irradiated as well as control plants. Height measurements of irradiated plants showed that they surpass the control plants (only irradiation with 5 c considerably retards the plant growth).

Observations on the beginning of flowering show that irradiation has a considerable effect on the onset and course of this phase of plants development. The effects of different doses are manifested differently on the flowering periods. With the irradiation of the late varieties Sterling and Osetinskaya at 0.5 c, the formation of tassels and flowering begins earlier, as does the development of female flowers. This is of great practical importance in the Latvian republic where early fall frosts significantly reduce the vegetative period of corn and do not allow an opportunity to get the cobs through the milky-dough maturity.

Table 4.73. Evaluation of the green mass of corn in fodder units

Treatment	Digestible nutrients per 100 parts of raw matter by weight				Starch equivalents in				The sum of starch equivalents	Reduction on digestibility	Residual sum of starch equivalents	Digestible units	
	proteins	fats	cellulose	NES	proteins	fats	cellulose	NES				per 100 parts of raw matter by weight	per 1 g
Control:													
stems	0.80	0.27	3.85	7.15	0.75	0.52	3.85	7.15	12.27	1.74	10.53	17.5	2,152
leaves	2.35	0.83	3.91	8.48	2.21	1.58	3.91	8.48	16.18	1.77	14.41	24.0	1,080
cobs	1.39	0.43	2.10	8.63	1.30	0.82	2.10	8.63	12.85	0.84	12.01	20.0	1,180
cob-sheaths	0.88	0.27	2.17	5.93	0.83	0.52	2.17	5.33	9.45	0.86	8.59	14.3	1,353
Total fodder unit													5,765
Total, %													100
Irradiation:													
stems	0.99	0.34	4.05	6.55	0.93	0.65	4.05	6.55	12.18	1.83	10.35	17.2	2,356
leaves	2.87	0.87	3.91	7.72	2.70	1.66	3.91	7.72	15.99	1.77	14.22	23.7	948
cobs	1.81	0.57	42	10.86	1.70	1.09	2.42	10.86	16.07	0.96	15.11	25.2	2,192
cob-sheaths	0.80	0.34	2.17	5.47	0 75	0.65	2.17	5.47	9.04	0.86	8.18	13.6	1,713
Total fodder unit													7,209
Total, %													125

Field experiments were conducted on the collective farms and state farms of the Latvian SSR. The trials have confirmed the possibility of increasing the yield of the green mass of corn with presowing irradiation of seeds. Large-scale commercial sowing done in different regions with different soil and climatic conditions showed the great effect of these conditions on the quantitative expression of the stimulation effect. When irradiated seeds were sown in the Riga region with heavy sandy-loam soils, the increase in the yield of green mass was considerably lower than when they were grown in the light, sandy loams of the Leipaiskii region (Table 4.74).

Histochemical research conducted on these corn varieties showed that when seeds were irradiated by ^{60}Co gamma-rays at 0.5–6 c, the quantity of chloroplasts in the leaves increases during mass flowering, but at 8 c it sharply drops. Along with the reduced quantity of chloroplasts, their morphology also changes and there is a different degree of agglutination. In doses which increase the quantity of chloroplasts, we observe a more intensive assimilation by the leaves.

Biochemical investigations on irradiated and unirradiated corn plants showed that under favorable soil and climatic conditions and with irradiation at 0.5–8 c, the sugar content increases considerably up to 2.37% at the leaf formation stage, but to 1–1.7% at the flowering stage in comparison with the control. Under the less favorable conditions of 1957, there was no increased sugar content in these varieties. In the late maturing corn varieties Sterling, VIR 25 and Osetinskaya, the content of dry matter in the leaves increased in comparison with the control with all doses of irradiation, but in the early maturing variety Voronezhskaya 76, it began to increase only with higher doses of irradiation—to the level of 6 and 8 c. The vitamin C content in the corn leaves increased by 2–12 mg% as a result of irradiation with 0.5–8 c.

The determination of the fodder values of the irradiated and control corn in large-scale commercial crops in the Latvian SSR showed that there were 5,300 fodder units per hectare in the green mass harvest of the unirradiated corn variety Sterling whereas, there were 6,625 fodder units in the harvest of irradiated corn. In the crop of the corn variety Voronezhskaya 76, the output of fodder units per hectare was 6,500 and 7,534 fodder units for those irradiated. Cytological investigations conducted on this material which had shown an increased mitotic index, indicated that when 8% mitosis was estimated in the irradiated cells, it was 4–6% in the control. The same phenomenon was noticed for other corn varieties as well.

In the Siberian botanic garden of the USSR Academy of Sciences [234, 235] biochemical investigations were conducted on the corn varieties Krasnodarskaya zubovidnaya belaya, Khar'kovskaya 23 and Dnepropetrovskaya 23, whose dry seeds were irradiated with doses of 0.1, 0.25, 0.5, 1 and 2 c. The effect of irradiation on the accumulation of ascorbic acid, carotine and

Table 4.74. Yield of the green mass of corn grown from seeds irradiated by ^{60}Co gamma-rays (1959)

Variety and place of research	Control, q/ha	Dose of irradiation, c					
		0.5		2		8	
		yield, q/ha	In percentage of control	yield, q/ha	In percentage of control	yield, q/ha	In percentage of control
Sterling							
Riga	365.7	430.0±0.41	117.3	447.7±0.32	122.4	282.1±0.12	77.5
Leipaiskii region	476.0	756.0±0.32	158.0	700.0±0.60	147.5	532.0±0.34	111.3
VIR 25							
Riga	417.8	418.01±0.31	104.1	451.2±0.25	108.0	380.0±0.19	91.0
Nereta	730.0	780.0±0.49	106.8	—	—	—	–
Leipaiskii region	392.0	476.0±0.57	121.4	504.0±0.29	138.5	448.0±0.74	114.2
Voronezhskaya 76							
Riga	421.7	492.5±0.26	116.8	457.8±0.36	108.8	339.5±0.03	80.0
Leipaiskii region	364.0	375.2	130.7	130.7±0.48	130.7	336.0±0.44	92.3

monosugars in the corn leaves was also studied. The data showed that the carotine content in the leaves of irradiated corn plants significantly exceeds the quantity of carotine in the leaves of the control plants. The maximum difference in carotine content in the leaves of experimental and control plants are given in Table 4.75.

An analogous picture with insignificant variations was observed in other varieties as well. Apart from carotine content, the content of ascorbic acid was also determined in the leaves of corn plants with the same doses of irradiation (Table 4.76).

An increase in the content of ascorbic acid appears to a large degree in the initial stages of development, but with the flowering stage it starts to decline. The authors noticed differences also in the content of monosugars in the leaves of control and irradiated plants, particularly at the early developmental stages of corn plants. For example, in the leaf formation phase, the content of monosugars in the leaves of irradiated plants exceeded the control by 34–40%.

The effect of the stimulation dose of irradiation on corn was also studied at the Institute of Plant Breeding and Genetics of the Academy of Sciences of the AzSSR [240]. The researchers irradiated air dried corn seeds of the variety Sterling with 0.5, 4 and 8 c and sowed them in the spring and fall to obtain two harvests of green mass in the year. Measurements of plant height showed that in Azerbaijan conditions the maximum increase in plant height was obtained with the irradiation dose of 4 c not 0.5 c. Tassel formation with 0.5 c began almost simultaneously with the control, whereas at 4 c tassel formation began six to seven days earlier. But irradiation with 8 c prolonged the vegetative period in the spring as well as the fall sowing (Table 4.77).

As evident from the table, the maximum yield increase was obtained through irradiation at 4 c (10.5–17.5%). Irradiation with 8 c in the spring sowing did not stunt the plants develop in conditions of the AzSSR, as was observed in the conditions in Moscow province and Latvian SSR.

This considerably higher stimulation dose for corn cultivation under conditions in Azerbaijan is a new and extremely interesting phenomenon noticed for the first time. We may presume that in conditions of very intensive insolation, the damaging effect of radiation is partially removed and the dose of 4 c, which is slightly suppressive, becomes the stimulation dose.

After analyzing the yield components of corn under conditions of AzSSR, it was established that one reason for the increased yield, as in the Moscow province also, is the increased number of cobs and their much earlier development. In the control, most plants had only one cob, whereas with irradiation at 4 c, a maximum number of plants had two cobs. In the control, 5% of the plants had three cobs, as did 18% among the irradiated plants.

Some biochemical investigations were done along with yield estimates on

Table 4.75. Effects of presowing irradiation on the synthesis of carotine in corn leaves (per 100 g of absolutely dry matter)

Dose of irradiation, c	Seedlings		Leaf formation		Flowering		Cob formation	
	Carotine content, mg%	In percentage of control	Carotine content, mg%	In percentage of control	Carotine, content, mg%	In percentage of control	Carotine content, mg%	In percentage of control
Control	48	—	66	—	47	—	40	—
0.10	66	137.5	82	124.3	52	110.6	41	102.5
0.25	70	145.8	92	147.0	59	125.5	50	125.0
0.50	65	135.4	90	136.3	65	138.3	52	130.2
1.00	69	143.8	70	106.6	65	168.3	52	130.0

Table 4.76. Effects of presowing irradiation on the content of ascorbic acid in the leaves of corn, mg% of absolutely dry matter

Dose of irradiation, c	Seedling emergence	Leaf formation	Tassel emergence	Flowering	Cob formation
Control	486.46	422.20	667.53	472.00	403.00
0.10	660.00	619.80	826.39	633.60	529.53
0.25	704.00	622.00	971.60	527.00	410.86
0.50	633.60	653.57	899.59	646.24	538.63
1.00	678.80	561.60	704.00	598.33	472.72

Table 4.77. Effects of presowing irradiation of corn seeds on the yield of green mass in spring and fall sowings in AzSSR

Dose of irradiation, c	Fall sowing		Spring sowing	
	Yield, q/ha	In percentage of the control	Yield, q/ha	In percentage of the control
Control	241.0	100.0	216.7	100.0
0.5	276.0	114.5	224.7	103.6
4.0	283.4	117.5	239.6	110.5
8.0	259.0	105.2	224.7	103.6

the green mass of corn with presowing irradiation. The ascorbic acid content and quantity of soluble sugars were determined in the leaves and cobs of irradiated corn grown in the spring and fall and at the different phases of plant development. Highest sugar content in the green mass was observed in the tassel formation phase in both control and irradiated plants, however, in the irradiated plants the sugar content exceeded the amount in the control plants through the entire vegetative period. The sugar content increased in the green mass of corn as a result of presowing irradiation, depending on the dose. The maximum increase was obtained through irradiation with 4 c; in the fall sowing, it was 30% and 25% in the spring sowing. Even in Azerbaijan with its shortened vegetative period, it was possible to get two crops in a year, even of such a late corn variety like Sterling due to the effects of irradiation.

Experiments on the presowing irradiation of corn were also conducted by N.G. Zhezhel', at the Leningrad Agricultural Institute [97, 102]. He irradiated seeds of the corn variety Sterling with doses of 0.25, 0.5, 1 and 1.5 c in the strength of 250 r/min. Field experiments were planned on the experimental farm of the Institute in Leningrad province. The total yield of green mass in the experimental plots exceeded the control by 9.6–13.5%. In this way, under field experiment conditions, the presowing irradiation of seeds with 0.5–1.5 c increased the yield potential of the plants. While conducting experiments with corn in pots, the increased yield of green mass through seed irradiation with 1 c was 9.5%.

At the Institute of Plant Physiology of the USSR Academy of Sciences, experiments were conducted for a number of years on the presowing irradiation of corn seeds to study the effect of irradiation on the grain yield and also in certain cases, on the yield of green mass. Dry seeds of corn were irradiated in the doses of 0.2–10 c. Doses up to 2 c activated the growth and development of plants by increasing the intensity of photosynthesis. Increased photosynthesis due to the effects of ionizing radiation is the result of the change in activity of the enzymatic systems and plastids. Physiological processes were activated not only by direct radiation effects, but also by subsequent metabolic processes in the irradiated plants.

An analysis of the ash of corn grown from irradiated seeds showed that, due to the effects of radiation, the selectivity of plants to microelements increases. All the noticed phenomena in plants grown from irradiated seeds increased the productivity of plants and improved the quality of the raw produce. This conclusion was made on the basis of hundreds of field experiments [65, 66, 69, 71, 74] conducted over ten years, in different agro-climatic zones of the Ukrainian SSR.

These data were the basis for tests on the presowing irradiation of corn seeds under conditions for commercial cultivation. In 1960, irradiated seeds of corn were sown in state varietal trials on collective farms of different provinces of the Ukraine. The data obtained are shown in Table 4.78. As may be seen from the data, in the majority of cases presowing irradiation led to increased yields of grain and green mass of corn.

In 1961, large-scale commercial tests on irradiated corn seeds were conducted at 17 centers of the Gossortoset (State varietal trial network) in various regions of the Ukraine. The stimulation effect appeared at some centers through the irradiation of seeds with 0.5 c, and with 1 c at other centers. These data showed more heterogeneity than the results of the experiment in 1960. Such heterogeneity can be explained by the fact that sowing in 1961 was done at more centers in different agro-climatic zones, which could have affected or changed the range of the stimulation doses. Besides, the seeds, brought for irradiation in open trucks to Moscow, remained in conditions of increased humidity and low temperatures before irradiation, which also may produce certain adverse effects on the uniformity of the stimulation effect. Moreover, in 15 of 17 cases, there was an increase in the yield of grain and green mass. The research conducted by the Institute of Plant Physiology of the USSR Academy of Sciences is of great

Table 4.78. Effects of irradiation of corn seeds of the variety VIR 25 by ^{60}Co gamma-rays on yield, q/ha (1960)

Treatment	Control	Dose of irradiation, c					
		0.5		1		5	
		Yield	Increase	Yield	Increase	Yield	Increase
Voznensenskii State Varietal Trial Center, Nikolaev province							
Grain at the complete maturity stage (moisture content 14%)*	34.2	36.9	2.7	37.4	3.2	37.7	3.5
Green mass at the milky-dough maturity stage of the grain**	208.0	218.0	10.0	212.0	4.0	210.0	2.0
Romenskii State Varietal Trial Center, Sumsk province							
Grain at the complete maturity stage	32.9	29.7	−3.2	30.3	−2.6	28.6	−4.3
Green mass at the milky-dough maturity stage of the grain	265.2	248.4	16.8	308.2	43.0	300.4	35.2
Collective farm of Man'kovskii region of Cherkass province							
Grain at the complete maturity stage	56.8	56.9	0.1	59.1	2.3	57.1	0.3
T.G. Shevchenko Collective farm of Prilukskii region of Chernigov province							
Grain at the complete maturity stage	17.4	18.5	1.1	20.5	3.1	17.6	0.2
Collective farm Ukraina of Novoselitskii region of Chernigov province							
Grain at the complete maturity stage	22.1	20.6	−1.5	26.0	3.9	22.5	0.4

*Error of the experiment 2.34%.
**Error of the experiment 1.72%.

significance for solving the problem on the positive effects of presowing irradiation. Variations in the optimal stimulation doses, depending on the cultivation conditions, show that it is essential to conduct studies in depth on the environmental factors influencing the effects of irradiation.

Presowing irradiation of air dried corn seeds of the variety Sterling was conducted by V.A. Guseva [80] at Gorky University to study the effects of low doses of irradiation on carbohydrate metabolism (Table 4.79). As is evident from the data in the table, the total quantity of carbohydrates in the leaves and stems of corn grown from irradiated seeds exceeds that of the control. With irradiation at 0.5 c, there is a much higher increase of sugar than with radiation at 1 c. Because of irradiation, mostly the water soluble sugars (mono and disaccharides) change and, to a lesser extent, the polycarbohydrates (starch and hemicellulose).

Table 4.79. Change of carbohydrate content in corn, due to the presowing irradiation of seeds by ^{60}Co gamma-rays (per 1 g of dry matter, mg)

Dose of irradiation, c	Leaves		Stems	
	8 July	19 July	8 July	19 July
	Mono + disaccharides			
Control	46.5	21.2	116.8	282.0
0.5	63.7	28.2	123.2	293.0
1.0	64.4	20.7	124.7	169.9
	Total quantity of polycarbohydrates			
Control	316.0	265.5	378.7	451.2
0.5	340.5	303.0	455.0	471.5
1.0	323.0	264.6	414.1	344.1

Seeds of the corn variety Sterling and some hybrids were irradiated in the agricultural institute at Kamenets-Podal'sk. In these experiments, the effects of x-ray irradiation with strength of 2.4–44 r/min (at 140–180 KV, strength of the current 10–20 mA, focal distance 50 cm) was compared with the effects of ^{60}Co gamma-rays with a strength of 41 r/min, in doses of 0.1–33 c. Three to five days after irradiation, three replications of seeds were sown in field plots with an area of 68.7 m^2 [238–154]*. X-rays in doses of 0.5–1 c and ^{60}Co gamma-rays in doses of 1–2 c produced the stimulation effect on the plant growth and the yield of green mass and cobs. The irradiated plants not only grew faster, flowered and bore cobs earlier but the ripening of cobs in them was set in 10–12 days earlier than in the control plants. Here, the green mass yield increased by 20–30% on an average and the cob yield by 15–20% in comparison with the control. Further increases in the dose led to decreases in the yield.

*So given in the Russian original—General Editor.

At the Kishinev Agricultural Institute [218], experiments were conducted to study the effect of different irradiation doses on the productivity of corn. Corn seeds of the variety Moldavanka Oranzhevaya were irradiated with 0.4, 0.6, 0.8, 1, 1.2, 1.4, 1.6, 1.8, 2, 2.5, 3, 3.5, 4, 4.5 and 5 c with different strengths of doses (160, 40 and 10 r/min). Maximum plant height was observed with irradiation in doses of 1.4–1.6 c with the strength of 160 r/min, with dose strength of 40 r/min—0.75–0.8 c and with dose strength of 10 r/min—0.4–0.45 c. The plants which received stimulation doses of irradiation (0.4–0.5 c) had strong stems, wide leaves and more height than that of the control. Also the flowering in these plants started two to three days earlier than in the control.

The researchers showed that low doses of irradiation cause stimulation, which extends to all developmental phases of corn and its productivity. In order to get the same effect with a much higher dose strength, we require much higher dose of irradiation and with low strength—low doses. It was shown by K.I. Sukach and E.D. Morozova that the presowing irradiation of corn seeds changes the duration and course of morphogenesis [219]. The period of cob and tassel formation is shortened with an increased dose. However, it is known that the longer the periods of formation the more productive and more vigorous the organ formation. On the contrary, organs formed in shorter periods are not well developed and are less productive. In these experiments, the lowest yield was obtained from the dose which caused faster rates of cob formation and the highest yield was obtained from the dose which caused a prolonged periods of cob formation, i.e., when flowering and cob formation began earlier than in the control. For example, the earliest flowering was observed with irradiation at 1 c and the dose strength at 40 r/min, but with irradiation at 0.5 c and the strength of the dose 10 r/min. With irradiation in these doses, the maximum yield increase varied within a range of 14–23% in comparison with the control.

In experiments on the presowing irradiation of corn seeds conducted at the Institute of Nuclear Physics of the UzSSR, through seeds irradiated at 0.5–1 c, the scientists noticed a 28% increase in the yield of cobs and much earlier plant development in comparison with the control. In plants grown from irradiated seeds, there was a tendency to produce more than one cob on each plant.

In their early experiments, Shull and Mitchell [350] subjected corn seeds of the variety Zubovidnaya yellow to x-ray irradiation at 100 KV voltage, and 5 mA current using a 1 mm thick aluminum filter and the dose strength about 38 r/min. The distance from the source to the object was 30 cm. The dose of irradiation was determined as per 1–5 minutes exposure time.

Before irradiation the seeds were soaked for 24 hours and one batch of seeds was then irradiated in the above conditions using the aluminum filter. Another batch was irradiated without the filter. On the fifth day, the

Table 4.80. Growth and development indices of corn grown from irradiated seeds (as per data of Shull and Mitchell)

Time of irradiation, min	Diameter of the stem		Roots		Stems			
			Raw mass		Dry mass		Raw mass	
	mm	%	g	%	g	%	g	%
Control	5.63	100.0	54.7	100.0	5.40	100.0	42.7	100.0
1	5.85	103.9	58.8	107.5	4.74	193.1	65.4	131.6
2	6.73	119.5	54.7	100.0	4.79	93.7	86.1	173.2
3	6.15	109.2	54.2	92.0	3.69	72.2	63.9	128.5
4	6.47	114.9	60.6	110.8	4.80	93.9	86.6	174.2
5	5.45	95.2	62.2	113.7	5.66	110.7	63.1	130.9

germination percentage of seeds irradiated with the filter was 84% and 72% of seeds irradiated without filter (germination percentage of the control—60%). In Petri dishes, the growth of coleoptiles in the irradiated seedlings was higher than the growth in the control. The dry mass of coleoptiles increased by 3–16% over the control. This shows that the irradiated corn seedlings used the reserve nutrients of the endosperm faster than the control (Table 4.80).

Thus data on the presowing irradiation of corn seeds obtained by Shull and Mitchell show the stimulation effect of certain irradiation doses on the growth and development of plants.

The positive results obtained by several researchers on the presowing irradiation of corn seeds with 0.5 c became the basis for conducting commercial trials on a large scale to determine the prospects of this measure and the possibility of its application in practical agriculture. The possibility of sowing irradiated seeds on large areas appeared in 1968, when D.A. Kaushanskii constructed the self-propelled, mobile gamma-unit Kolos with a huge capacity to irradiate 1 ton of seeds per hour at an integral dose of 0.5 c [125].

Over three years [152] 230 tons of seeds of 19 different crops grown in the MSSR were irradiated on the Kolos unit and sown in an area of about 15,000 hectares, of which 1,445 hectares were under corn. The irradiated seeds were sown on the best 95 state farms, collective farms and in plots of the research institutions of the MSSR. After three years of trials (1971–1972) [sic], the sowing of irradiated seeds was continued with further expansion of the crop areas, estimated to be about 50,000 ha in 1972.

The location of farms where large-scale commercial trials on the presowing irradiation of corn seeds (Table 4.81) on the Kolos gamma-unit were conducted, are shown in the map of the MSSR (Fig. 4.8).

The use of the Kolos unit provides the farms with a profit of more than one thousand rubles per hour. Hence, in the event of the full utilization of

Table 4.81. Results of the presowing irradiation of corn seeds in the dose of 0.5 c in the MSSR

Year	Yield		In percentage of the control
	Control plants, q/ha	Irradiated plants, q/ha	
1968	44.6	49.6	111.0
1969	36.2	39.1	108.1
1970	33.9	38.4	113.3
1971	28.4	32.9	115.8
Average			112.0

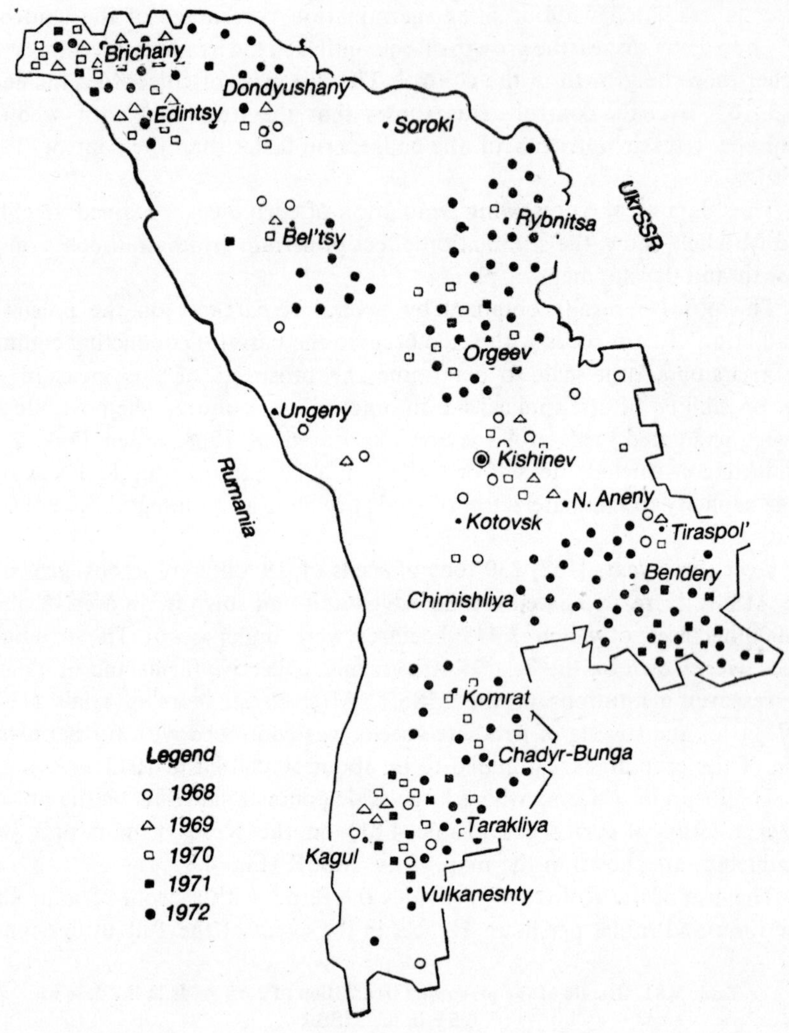

Fig. 4.8. Location of farms where large-scale commercial sowing with irradiated seeds was done in the MSSR.

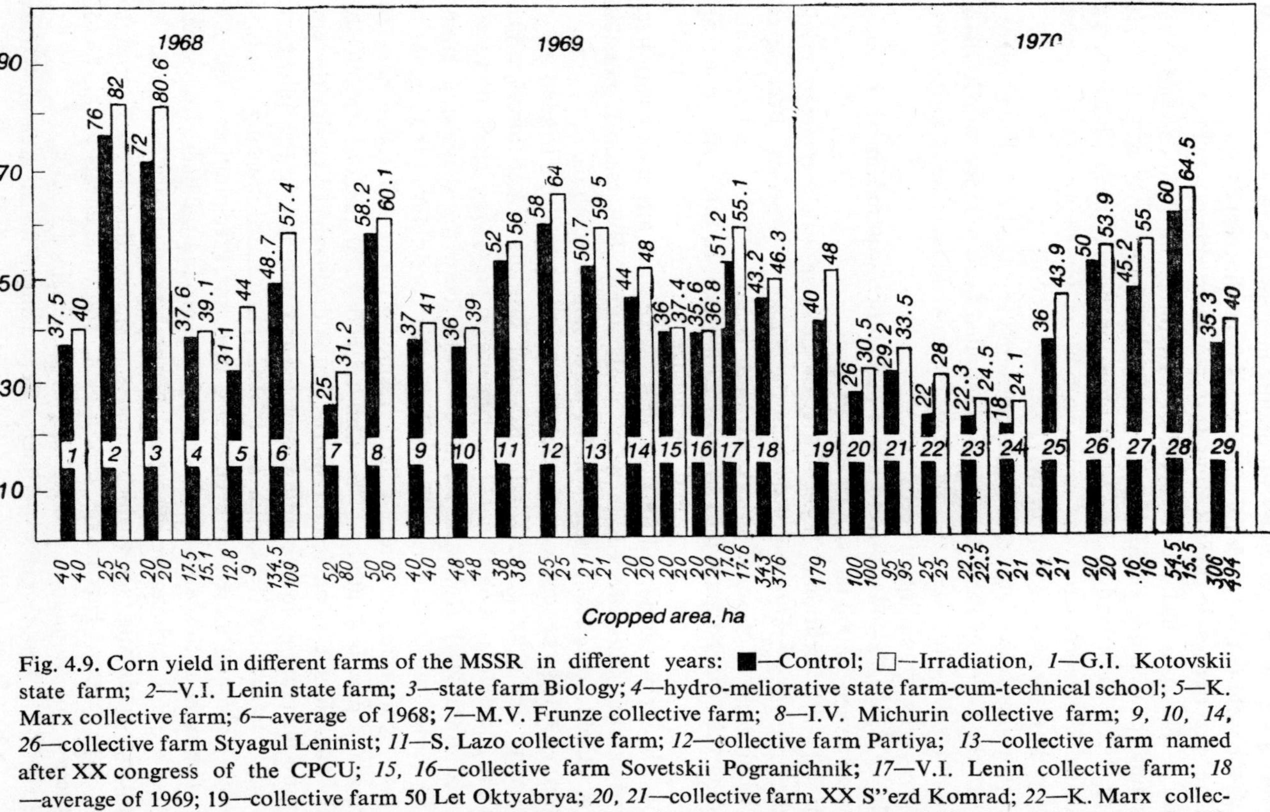

Fig. 4.9. Corn yield in different farms of the MSSR in different years: ■—Control; □—Irradiation, *1*—G.I. Kotovskii state farm; *2*—V.I. Lenin state farm; *3*—state farm Biology; *4*—hydro-meliorative state farm-cum-technical school; *5*—K. Marx collective farm; *6*—average of 1968; *7*—M.V. Frunze collective farm; *8*—I.V. Michurin collective farm; *9, 10, 14, 26*—collective farm Styagul Leninist; *11*—S. Lazo collective farm; *12*—collective farm Partiya; *13*—collective farm named after XX congress of the CPCU; *15, 16*—collective farm Sovetskii Pogranichnik; *17*—V.I. Lenin collective farm; *18*—average of 1969; 19—collective farm 50 Let Oktyabrya; *20, 21*—collective farm XX S''ezd Komrad; *22*—K. Marx collective farm; *23, 24, 25*—collective farm XX S''ezd MANR; *27*—collective farm named after XXI Congress of the CPCU; *28*—collective farm Vyatsa Noue; *29*—average of 1970.

the capacity of the machine during sowing time, it not only completely recovers its cost, but provides an additional profit [26, 151, 152]. Fig. 4.9 shows the yield data of corn for three years from different farms.

The experiences in MSSR found response in other republics of the country. The provincial directorate of agriculture of Pavlodar province, following the example of Moldavia, acquired a Kolos unit and, in 1971, sowed 7,000 ha of land in the 12 best farms of the province with irradiated seeds including corn. Corn was sown on 540 ha for silage and the green mass yield exceeded the control by 11%.

According to the data from Yu.A. Martem'yanov and A.V. Kalashnikov, in 1971 the farms of Pavlodar province supplied farm produce worth 104.9 thousand rubles to the state by virtue of the yield increase resulting from the presowing irradiation of seeds.

In 1972, the Pavlodar directorate of agriculture acquired another Kolos unit and sowed 20,000 ha with irradiated seeds.

It is worth mentioning a peculiarity of the large-scale commercial trials on the presowing irradiation of seeds in the Pavlodar province. Because of the wind erosion of soils, strip cultivation is done there on a large scale, dividing the cropped areas into strips of 25 ha each. In this system one strip is sown with the main crops and the adjacent strips with soil protecting crops. The strips earmarked for main crops were sown alternately with irradiated and unirradiated seeds in many replications. This made it possible to get suitable data from each farm for mathematical analysis.

Since 1971, large-scale commercial trials on the presowing irradiation of seeds were conducted at the Kirghiz Research Institute of Agriculture, which also had acquired a 'Kolos' unit. Experiments on presowing irradiation were planned in the Issyk-Kul basin for three crops: corn for silage and grain, spring barley and cucumber. The soil-climatic conditions of the regions in the Issyk-Kul basin are unique and the results of the experiments conducted there are of practical interest. The basin is 1,600–4,200 m above mean sea level. The climate is moderately warm, but is distinguished by drastic temperature fluctuations in late spring, summer and early fall. In the hilly part, the day temperature is as high as +30°C, but toward morning it falls to around +3°C and sometimes even to 0°C or below. Precipitation is 260–350 mm. The soils of this zone have increased background radioactivity because of the natural radioactive elements. Natural radioactivity is almost an unstudied, modifying factor which imparts a special theoretical interest to these experiments. This makes it possible to study the effects of this factor on the level of stimulation and stability of the obtained results.

In 1971, 17 commercial experiments were planned for an area of 400 ha in the 15 best farms of the republic. For growing irradiated corn seeds for grain purpose, irradiation doses of 0.5, 1.5, 2.5 and 3 c were studied on the varieties Sterling, Krasnodarskaya 4 and Krasnodarskaya 5. An increase in the yield of the green mass, through irradiation with 1.5 c, was 8–25%, while it was 2.8–50% at 3 c. An analysis of the yield components shows that the yield of silage mass increases mainly by virtue of the increased number of cobs per plant, as in other soil-climatic zones. For example, in the Dzhingi-Pakhta state farm nine of 100 plants in the control had two cobs, but with irradiation in the dose of 0.5 c, there were 15 such plants and 18 with 1.5 c. As a result of irradiation the number of cobs with milky-dough maturity also increased. It was 12.1% in control, and 45.1% for the irradiated plants on the state farm Dzhingi-Pakhta and 52% at the Kirghiz breeding station.

When determining the fodder values of the green mass of corn, a tendency toward increased carotine and calcium contents was detected. In the green mass grown from unirradiated seed, the carotine content was 11.30 mg/kg and 4.67 mg/kg of the fodder and calcium, with irradiation in the dose of 0.5 c, the carotine content increased to 23.5 mg/kg (or to 206%) and the calcium content to 5.01 mg/kg (or to 107.3%). At the dose of 1.5 c, these indices were 23 mg/kg (203.6%) and 5.80 mg/kg (124.2%) respectively. These results are new and in the event of good reproducibility in subsequent experiments they can be of great practical importance.

According to data from the experimental farm of the Kirghiz Research Institute of Agriculture, presowing irradiation increases the content of digestible proteins in the grain: in unirradiated corn, their content was 36.85% per kg, with irradiation at 0.5 c it was 65.10%/kg and 52.50%/kg at 1.5 c. Stimulation doses of 0.5–1.5 c were established for corn and it was planned to continue large-scale commercial trials.

There are also some data about the presowing irradiation of corn from abroad. In the report from MacKey, et al., at the Geneva conference on the peaceful utilization of atomic energy in 1971, results were presented of four years experiments on the presowing irradiation of corn seeds in the doses of 0.1–4 c. According to his data, the yield increased 40% in comparison with the control.

The Italian radiobiologist Coldera conducted a series of experiments on the irradiation of corn seeds of the variety Saturno Tv. 37 R. Seeds of this variety were of Italian reproduction and had 99% purity and 93% germination. The plot size was 20.25 m^2, six replications in all were made, and 60 plots were used for a total area of 1,215 m^2. The plots were arranged in a random block design. Results from these experiments are summed up in Table 4.82. The data were processed mathematically which testifies to their reliability and significance.

Presowing irradiation of corn seeds of the variety Mv-1 in the dose of 0.5 Krad was done in Hungary. Seeds were sown in two replications in plots measuring 0.57 ha. Results of this experiment are shown in Tables 4.83 and 4.84.

Table 4.82. Results of field experiments on the irradiation of the corn variety Saturno Tv. 37 R

Dose of irradiation, c	Weight of green mass per plot, with 15.5% moisture, kg	Weight of grain per plant, g	Number of plants	In percentage of the control
Control	115.8	189	611	100
0.1	124.8	209	599	111
0.2	120.7	201	600	107
0.3	121.2	203	597	108
0.4	121.0	195	607	104
0.5	121.8	196	619	104
0.6	121.5	197	616	105
0.8	121.1	200	603	100
1.0	122.1	200	607	106
1.2	120.9	196	612	105

Table 4.83. Effects of gamma-rays on the growth and development of the corn variety Mv-1

Dose of irradiation, Krad	Plant height, cm	Total mass of plants, g	Mass of cobs, g	Number of leaves
Control	147.2±2.5	708.7±27.3	304.7±8.6	10.70±0.14
0.5	157.1±2.3	769.7±27.4	344.7±12.5	11.38±0.27
Increase of yield:				
g	9.9	61	40	0.68
%	106.9	108.61	131.28	2.99

Table 4.84. Results of field experiments on the presowing irradiation of corn seeds of the variety Mv-1

Dose of irradiation, Krad	Mass of plants from one plot, kg			In percentage of the control
	1st replication	2nd replication	Average	
Control	32.5	28.3	30.40	100
0.5	37.2	30.1	33.65	110.69

The data obtained in Hungary are closer to our data regarding the value of the optimal stimulation dose as well as the value of the increase in yield. Results of the above described experiments are shown in Table 4.85.

Buckwheat. Experiments on the study of the effects of irradiation on

Table 4.85. Consolidated data on the presowing irradiation of corn seeds

Place of research	Year	Researcher	Dose of irradiation	Increase of yield in comparison with the control	Biochemical changes
1	2	3	4	5	6
Institute of Plant Physiology of the Academy of Sciences of the UkrSSR	1960	P.A. Vlasyuk	0.5–5.0 c	Green mass, 2–25 q/ha Grain, 0.2–3.5 q/ha	—
Leningrad Agricultural Institute	1959	N.G. Zhezel'	0.5–1.5 c	Cobs, 26–84%	—
	1960		1.0 c	Green mass, 9.5%	—
	1960		0.5–1.5 c	Green mass, 9.6–13.5%	—
Institute of Biology of the Academy of Sciences of Latvian SSR	1959 Riga	K.K. Roze, V.T. Kietse	0.5–2.0 c	Green mass, 17.3–22.4%	—
	1960 Liepaiskii region		0.5–2.0 c	Green mass, 25.8–47.5%	—
Institute of Biology of the Academy of Sciences of Latvian SSR		K.K. Roze, V.T. Kietse	0.5–2.0 c	Green mass, 21.4–38.5%	—
	1960 Riga	V.T. Kietse	0.5–2.0 c	Green mass, 0.2–8%	—
			0.5–2.0 c	16.8–8.8%	—
			0.5–2.0 c	30.7%	—
Botanic Garden of the West Siberian branch of the USSR Academy of Sciences	1958	V.S. Feodorova	0.5 c	—	Increase in the content of vitamin C by 2.2–33.6 mg %, monosugars by 34–40%

(Contd.)

Table 4.85 (*Contd.*)

1	2	3	4	5	6
Botanic Garden of the West Siberian branch of the USSR Academy of Sciences	1962	N.D. Sidorenko	0.5–1.0 c 0.5–2.0 c 0.5–1.0 c 0.5–1.0 c	Grain, 24.6–27.6% Grain, 17.5–18.7% Green mass, 4.7–24.5% Cobs, 9.5–10.7%	— — — —
Institute of Biophysics of the USSR Academy of Sciences	1958–1960	N.M. Berezina	0.5 c	Green mass, 8–18%	—
Institute of Biophysics of the USSR Academy of Sciences	1961 Sterling	N.M. Berezina	0.5 c	21%	Increase in fodder values by 25%
	1962 Sterling	R.R. Riza-Zade	0.5 c	Green mass, 8–17%	—
Institute of Biophysics of the USSR Academy of Sciences	1963 Sterling	N.M. Berezina R.R. Riza-Zade	0.5 c	Green mass, 15.3%	Increase in fodder values by 18.5%
	1963 Hybrid Bukovin-skii-3		0.5 c	22.7%	Increase in fodder values by 26.5%
Agrophysics Institute	1961 Sterling	N.F. Batygin	0.5 c	Green mass, 17–23%	—
Gorky State University	1961	V.A. Guseva	0.5 c	Green mass, 17%	Increase of carbohydrate content
Institute of Nuclear Physics of the Academy of Sciences of the UzSSR	1962	S.A. Anastasov	0.5 c	Green mass, 28%	—

Institute of Genetics of the Academy of Sciences of the AzSSR	1960	A.I. Khudadatov	4 c	Green mass, 14.5%	—
	1961 Spring sowing		4 c	Green mass, 17.5%	—
	1961 Fall sowing		4 c	Green mass, 12%	Increase in vitamin C content by 7.9 mg%
Kishinev Agricultural Institute	1964	K.I. Sukach,	1 c	32%	—
	1968	V.N. Lysikov,	0.5 c	Grain, 13.1%	—
	1968	G.Ya. Rud',	0.5 c	Green mass, 8%	—
	1969	K.I. Sukach,	0.5 c	Grain, 8.1%	—
	1969	N.M. Berezina,	0.5 c	Green mass, 19.6%	—
	1970	D.A. Kaushanskii	0.5 c	Grain, 13.6%	—
	1970		0.5 c		—
KazSSR	1971	A.V. Kalashnikov, Yu.A. Martem'yanov	0.5 c	Green mass, 11%	—
	1972		0.5 c	Green mass, 16.9%	Enhancement of maturity by 5–6 days
Canada	1968–1971	MacKey	0.1–1.0 Krad	Grain, 34%	—
Hungary	1968	Simon, Menyhert, Pannonhalmi	0.5 Krad	Green mass, 10.6%	—
			0.1 Krad	Grain, 11%	—
Italy	1971	Coldera	0.1–0.3 Krad	7–11%	—

buckwheat seeds are represented by the solitary research conducted during past years in the USSR. The effects of irradiation in this research were considered not only on the basis of the change of growth, development and yield of plants, but also on the basis of the biochemical changes.

V.A. Guseva [80] conducted two series of experiments on buckwheat. In the first series, she studied different versions of the presowing irradiation of air dried seeds and in the second series, the effects of the chronic gamma-irradiation of growing buckwheat in a gamma-field, where a ^{60}Co isotope with the activity of 1 c was the source of radiation.

In the first series of experiments, the presowing irradiation of seeds was done with doses of 0.5 and 1 c, with the strength of 720 r/min. The seeds were first class with a conditioned moisture content and from the harvest of the previous year. Along with the grain yield estimate, the change in the carbohydrate and protein metabolism in irradiated seeds was also taken into consideration (Table 4.86).

As is evident from the data presented above, the presowing irradiation of buckwheat seeds in doses of 0.5 and 1 c increased the content of carbohydrates insignificantly and to a considerably greater extent, increased the content of proteins, particularly with irradiations of 0.5 c. In this dose, the maximum increase of grain yield was also observed and exceeded the yield of the control by 59% in 1962.

Table 4.86. Effects of the presowing irradiation of seeds by gamma-rays on the yield of buckwheat and its quality

Dose of irradiation, c	1961		1963			
	Carbohydrates, %	Proteins, %	Carbohydrates, %	Proteins, %	Grain yield, q/ha	Grain yield in comparison with the control, %
Control	66.0	14.3	71.5	5.9	4.2	100.0
0.5	67.1	18.5	77.7	8.4	6.7	159.6
1.0	67.1	16.0	71.5	7.6	4.3	103.0

Apart from the total nitrogen content in the irradiated buckwheat seeds, the effect of irradiation on nucleic metabolism was also determined. For this purpose the seeds were irradiated with 0.3, 0.5 and 10 c. The increased nucleic acid content in the irradiated plants occurred as a result of the increased RNA content, while the quantity of DNA was close to or even below that of the control. The maximum increase in the nucleic acid content was observed with irradiation at 0.3 c. The nucleic acids were distributed unevenly in the stems: they were almost two times more in the upper part than in the lower part.

Observations were also conducted on changes in the activity of the respiratory enzymes, viz., peroxidase and polyphenoloxidase in the aerial part of the seedlings and in the roots. Peroxidase was most active in the roots of the buckwheat seedlings grown from seeds irradiated with 0.5 c and somewhat less active in roots with the dose of 10 c. In the aerial part of the seedlings however, there were no considerable differences in the activity of this enzyme. Polyphenoloxidase activity in the roots of irradiated seedlings was less than in the control. The irradiation of seeds with 10 c reduced the activity of almost all enzymes under study, with the exception of ascorbinoxidase.

Increased respiratory enzyme activity intensifies the oxidizing enzymatic reactions in germinating the irradiated seeds. This causes a much faster flow of nutrients toward the embryos of irradiated seeds, accelerates their germination and causes much faster seedling development. It can thus be considered one factor causing the stimulation effect.

In 1971, in the Pavlodar province of the KazSSR, a large-scale commercial trial (in two replications) on the presowing irradiation of buckwheat seeds of the variety Bogatyr was conducted on an area of 30 ha. Primarily, an optimal dose of 0.4 c was established for irradiating buckwheat seeds [158]. The grain yield in the control was 4.5 q/ha and 6.5 q/ha with irradiation i.e., the yield increase was to the tune of 44% in comparison with the control.

Consolidated data from experiments on buckwheat are shown in Table 4.87.

Rice. The Japanese scientist Jamado, in 1917, conducted experiments on presowing irradiation of rice seeds which he soaked in saline water for seven days and then irradiated with 3, 5, 7 and 10 NED and soaked them again in the water. With seed irradiation at 3 NED in field experiments, he noticed a 40% increase in grain yield but irradiation with 7 and 10 NED stunted plant development and reduced their yield by 2.4 and 5.4% respectively [298].

In 1919, Komuro irradiated dry and soaked rice seeds. In experiments the optimal stimulation doses were 5 and 10 NED. Dry seeds were more radiosensitive than soaked ones. The damaging effects of irradiation increased proportionately with an increase in moisture content in the seeds. Average doses stimulated the growth and development of rice plants [310, 311, 313, 314].

In this way, despite great significance of rice for the national economy, at present we have only a limited number of old reports which do not enable us to judge the possible effects of stimulation on this crop.

Table 4.87. Consolidated data on the presowing irradiation of buckwheat seeds

Place of research	Year	Researcher	Dose of irradiation, c	Increase of yield in comparison with the control, %	Change of biochemical composition	Enhancement of development
Gorky State University	1955–1957	V.AG. useva	Chronic irradiation	53.4–42.3	Increase of protein content by 42–53%	—
	1957–1963		0.3–10	61	—	—
Institute of Biophysics of the USSR Academy of Sciences	1954–1957	L.P. Breslavets, N.M. Berezina	Chronic irradiation	38–42	Increase in the content of rutin by 18%	Enhancement of flowering
Pavlodar Provincial Directorate of Agriculture	1971	A.V. Kalashnikov, Yu.A. Martem'yanov	0.4	44	—	—

PERENNIAL GRASSES

Clover (*Alfalfa*). Experiments on the presowing irradiation of clover seeds by ^{60}Co gamma-rays were conducted at the All-Union Research Institute of Fertilizers and Agricultural Soil Science [130, 131]. The authors studied the effect of gamma-rays on the quantity and quality of the yields of first and second generations, as well as the effects of recurrent irradiation. Seeds of single-cutting and double-cutting clover were used. The doses of irradiation on air dried seeds were 0.1–1 c. Plants from the irradiated seeds were grown in pots in vegetative experiments using heavy loam and sod-podzolic soils, as well as in small plot field experiments in six replications. Data about the yield of the vegetative mass of clover for two years are presented in Tables 4.88 and 4.89.

These data show that the presowing irradiation of clover seeds with 0.5 and 1 c increased the yield of clover heads two to three times and almost did not affect the weight of aerial vegetative mass. The optimal dose for single-cutting clover was 1 c and for double-cutting clover—0.6 c. An increase in the yield of reproductive organs occurred by virtue of their increased number and not by absolute mass.

No increase in the germination of irradiated clover seeds and their germination energy was observed. According to the data from A.S. Palamarchuk [178], the irradiation of air dried seeds with doses of 0.25–1 c stimulated the growth and development of clover. It is interesting to note that

Table 4.88. Effects of the presowing irradiation of seeds on the yield of aerial mass of clover

Dose of irradiation, c	Yield of heads		Yield of vegetative mass	
	g/pot	%	g/pot	%
		1957		
Control	2.8±2.4	100	41.0±2.4	100
0.5	6 0±2.3	214	42.3±4.7	103
1.0	6.8±1.4	240	44.9±3.6	109
2.0	2.9±1.5	103	40.2±5.5	99
		1958		
Control	0.3±0.1	100	47.8±2.4	100
0.5	0.7±0.2	220	45.2±4.1	95
0.6	1.1±0.5	320	52.9±2.0	110
0.8	0.6±0.5	175	45.5±2.1	94
1.0	0.4±0.2	136	44.4±0.6	95
1.45	0.6±6.3	190	44.9±1.3	94

Table 4.89. Effects of the presowing irradiation of seeds on the quantity and mass of clover heads

Dose of irradiation, c	Number of heads per pot	Average mass of one head, g
	1957	
Control	18	0.16
0.5	38	0.18
	1958	
Control	7	0.05
0.5	11	0.07
0.6	16	0.06

irradiating clover seeds with much higher doses 7.5–15 c, suppressed the vegetative growth in the first year, but stimulated the growth and development of the plants in the third year. Storage of the irradiated seeds under favorable conditions and their subsequent sowing on peat soil made it possible to get a 33% increased yield of the green mass of clover in comparison with the control. A combined application of irradiation and gibberelline increased the yield of plants and the protein content in leaves by 1.8%.

Köernike [307], in much later experiments, tried the effect of different doses of irradiation on clover. He irradiated dry and soaked seeds of different agricultural plants including clover in different doses varying from low to lethal. Köernike, writing about these experiments, said that the effect of radiation on the dry and soaked seeds is governed by the Arndt-Schultz law, i.e., low doses accelerate the growth and development of plants grown from irradiated seeds, but much higher doses suppress. We must pay attention to Köernike's conclusion because in his earlier works he negated the possibility of the stimulation effect of presowing irradiation prevailing during the entire vegetative period and recognized the effect only at initial stages of development of the irradiated seeds.

Research workers of the Institute of Biophysics of the USSR Academy of Sciences [173] showed that with the presowing irradiation of clover seeds in the dose of 1 c, in laboratory experiments, in field and in pot culture experiments, the yield of the green mass of clover increased by 8–12%. Besides, the researchers determined the content of cyanides in clover plants and established a direct correlation between cyanide content and the radiosensitivity of clover, whose seeds were found to contain cyanogenic glycosides. The cyanide content in such valuable fodder crops like lucerne, clover and sorghum is a negative index. The application of mineral and organic fertilizers facilitated not only the increased yield of these crops, but also invariably increased the hydrocyanic acid content in them. With the gamma-irradiation of air dried seeds, the content of cyanides was reduced in the plants, but the yield of the green mass increased.

In the Azerbaijan Research Institute of Agriculture [202], three years of trials were concluded on the presowing irradiation of multi-cutting clover and observations about the yield of green mass and qualitative changes of the plants were taken into consideration. The trials showed an increase in the contents of fat, cellulose, proteins, ashes and nitrogen-free extractive substances in different organs of the plants. All estimates were made on five cuttings. The increased hay yield in different years varied within a range of 15–20% in comparison with the control.

The presowing irradiation of lucerne seeds of the variety Tashkentskaya was done at the All-Union Research Institute of Cotton Cultivation [172] in a wide range of doses—5–270 c, with the strength of 171 r/sec. In the second cutting, the hay yield of lucerne exceeded the control by 16–20% as a result of irradiating seeds with 2–40 c.

In the second year, as a result of irradiating seeds with 20–30 c, the hay yield exceeded the control by 10–34 q/ha, i.e., by 4–14%.

In another experiment conducted by this researcher, the optimal dose was also found to be very high—40 c and in this dose the increased hay yield for four cuttings was 31%. This experiment shows the very high radioresistance of lucerne which, even with irradiation at 100 c, produced green mass only 15% less than the control.

These data have been confirmed by the work of A. Nigmanov, who established the high radioresistance of such plants, including lucerne, whose seeds contain cyanogenic glycosides. According to the data from V.L. Mukhanova, in lucerne hay grown from irradiated seeds, a much higher nitrogen content was observed, which improved the fodder qualities of hay.

Results of experiments on perennial grasses are presented in Table 4.90.

FIBER CROPS

Flax. Flax fiber seeds were irradiated at the Institute of Biophysics of the USSR Academy of Sciences, jointly with the All-Union Research Institute of Flax Cultivation [230] using gamma-rays in doses of 0.2, 0.4, 1 and 2 c. Two flax varieties, viz., I-7 and L-1120 which are distinguished from each other by their physiological and technological properties were used in the experiments. The flax variety I-7 has a vegetative period of 79 days, its fiber is of average quality with number 16–18, which is used to prepare fine linen cloth. Seeds are small and they contain 35–37% oil. The flax variety L-1120 is a late variety with a vegetative period of 89 days. Fibers of this variety are longer and thicker (number 12–14) than fibers of the variety I-7 and they are used to prepare coarse linen cloth. This variety has much larger seeds and contains 40–41% oil.

Table 4.90. Consolidated data on the presowing irradiation of clover and lucerne seeds

Place of research	Year crop	Researcher	Dose of irradiation, c	Increase of yield in comparison with the control, %	Change of biochemical composition	Enhancement of development
All-Union Research Institute of Fertilizers and Agricultural Soil Science	1957–1959 clover	O.K. Kedrov-Zikhman, N.I. Borisova	0.2–1.0	Green mass, 3–10	—	—
Institute of Plant Physiology of the Academy of Sciences of the UkrSSR	Clover	A.S. Palamarchuk	0.5–1.0	Green mass, 33	1.8% increase in the protein content in green mass	—
Germany	1914–1920	Köernike	—	—	—	Stimulation of growth and development of plants
Institute of Biophysics of the USSR Academy of Sciences	1960	N.M. Berezina	0.5	Length of roots, 32, length of aerial parts 28	—	—
Institute of Biophysics of the USSR Acadmy of Sciences	1969–1972 lucerne	A. Nigmanov	1	7–10	Decrease in the content of hydrocyanic acid	—
All-Union Research Institute of Cotton Cultivation	1972 lucerne	V.L. Mukhanova	1	16–20	—	—
Research Institute of Agriculture of the AzSSR	1970–1972 lucerne	R.R. Riza-Zade	1	15–20	—	—

Field experiments were conducted in ten replications on small plots of the farms of the All-Union Research Institute of Flax at Torzhok. Seeds were sown ten days after irradiation. Seeds irradiated in doses of 1 and 2 c produced more uniform seedlings with their energy of germination above 60% as against 40% in the control. From irradiated seeds, the seedling emergence was completed within seven days, whereas in control this period was prolonged to nine days. Measurements on stem lengths before harvesting are shown in Table 4.91 [142].

The optimal stimulation doses to get the longest fiber were 1 and 2 c.

Apart from fiber length, its quality vis-a-vis different doses of irradiation was also determined. Fiber quality isd etermined by its fineness, strength and elasticity and it is designated by a corresponding number—the higher the number, the more valuable the fiber. The data on the variety I-7 with much higher fiber quality are of special interest. The fiber with the number 16 was obtained in the control, whereas after seed irradiation with 2 c, this variety produced fiber with number 26. As far as the variety L-1120 was concerned, its control as well as irradiated seeds produced fiber of the same quality—number 14. However, the quantity of fiber with all doses of irradiation was somewhat higher than in the control.

Table 4.91. Length of the flax stems with the presowing irradiation of seeds by gamma-rays

Dose of irradiation, c	I-7		L-1120	
	Total	Technical	Total	Technical
Control	74.4	68.8	79.7	72.0
0.2	76.7	70.0	84.0	75.4
0.4	76.2	67.7	82.0	78.0
1.0	76.6	70.0	89.1	77.3
2.0	78.4	70.0	83.0	75.8

Oil content was determined in the flax seeds as well. The irradiated seeds of the variety I-7 showed no deviations from the control regarding oil content. With the irradiation of seeds of the variety L-1120 with 1 c, an increase of 0.7% in the oil content was observed in comparison with the control.

In the Agricultural Academy named after K.A. Timiryazev [237], the stimulation of the growth and development of plants was noticed as a result of the irradiation of flax seeds with x-rays.

Presowing irradiation of linseeds was done at the All-Union Research Institute of Oil Seed Crops. With the dose of 1 c, there was a 22% increase of seed yield in comparison with the control.

Results of experiments on flax seeds are shown in Table 4.92.

Table 4.92. Consolidated data on the presowing irradiation of flax seeds

Place of research	Year	Researcher	Dose of irradiation, c	Increase of yield	Change of biochemical composition and technological indices
K.A. Timiryazev Agricultural Academy	1935	G. Frolov	1	Increase of energy of seed germination and stimulation of the growth of seedlings	—
Institute of Biophysics of the USSR Academy of Sciences	1961–1962	A.M. Kuzin, L.M. Kryukova	1–2	—	Variety I-7: Increase of fiber length by 2.2–4 cm. Variety L-1120: Increase of oil content in seeds by 0.7%
All-Union Research Institute of Oil Seed Crops	1964	D.P. Umen	1	Increase of yield by 22%	Increase of fiber number

Ambari-hemp, Jute, Dogbane. At the Institute of Nuclear Physics of the Academy of Sciences of the UzSSR, in collaboration with the department of Industrial Crops of the All-Union Research Institute of Cotton Cultivation [10], experiments were conducted to study the effects of ionizing radiation on the growth and development of ambari-hemp variety 1500 and jute variety Uzbekskii 53—the most valuable bast fiber crops which have been long cultivated in Uzbekistan. Air dried seeds were irradiated by gamma-rays in doses of 1,000–100,000 rep (physical roentgen equivalent) and by thermal neutrons. Phenological observations show the very high radioresistance of ambari-hemp and jute in general, and more particularly, the ambari-hemp which exceeded the control with respect to several indices, as a result of irradiation in the dose of 35,000 rep. Irradiation in doses of 1,000–35,000 rep facilitated the increase of germination, energy of seed germination, increase of height and diameter of the stems, which all contributed toward an increased fiber yield by 12.8–46.2% (Table 4.93). Indices contributing toward an increased fiber yield are the density of plant stand per unit area, plant height and diameter of the stems. With the growth of these indices there is a natural increase in the yield of the technical or industrial fibers which are contained in the stems of the plants. The technological indices of ambari-hemp and jute fibers did not change as a result of irradiation, however, doses much higher than those shown in Table 4.94 suppressed the growth and development of plants.

Air dried seeds of these plants were also irradiated by slow neutrons in the canal of the reactor with the density of the nuetron flux at 1.6×10^{10} neutrons/($cm^2 \cdot sec$). The irradiation of seeds in this flux for 5, 15 and 55 minutes, as well as irradiation by gamma-rays, increased the germination percentage and energy of seed germination, which facilitated a higher density of plant stands. In contrast with gamma-irradiation, irradiation by neutrons had less effect on the changes in the diameter of the stems. Results of phenological observations and yield estimates for seeds irradiated by neutrons are shown in Table 4.94.

As a result of the presowing irradiation of seeds of ambari-hemp and jute, there was an additional branching of plants which produced two and sometimes even three equally valuable stems.

Zaurov conducted experiments on irradiation of the air dried dogbane seeds of the variety Amudar'inskii. The observations showed that irradiation caused insignificant increases of plant height, but with the dose of 1 c there was an increase in their tillering.

Table 4.95 shows the results of experiments on the above crops.

Cotton. The presowing irradiation of such a valuable crop as cotton naturally drew the attention of several researchers because enhanced and increased yields of raw cotton are very important to the national economy. The possibility of increasing the yield of raw cotton through irradiation is

Table 4.93. Biological peculiarities of the growth and development of ambari-hemp and jute after the presowing irradiation of seeds by ^{60}Co gamma-rays

Dose of irradiation, fer	Energy of seed germination, %	Germination, %	Density of plant stand, thousand/ha		Morphology		Yield in comparison with the control, %	
			after seedling emergence	before harvesting	plant height before harvesting, cm	diameter of stem, cm	stems	fiber
Ambari-hemp								
Control	79.2	91.5	58	45	239	7.8	100.0	100.0
1,000	79.7	91.0	58	49	245	8.5	140.3	146.2
15,000	82.0	94.5	56	49	242	8.1	102.2	102.5
35,000	84.2	93.5	54	46	230	8.7	111.6	112.8
Jute								
Control	92.0	94.2	162	69	205	7.8	100.0	100.0
1,000	95.7	96.0	162	85	213	8.4	108.8	120.5
7,000	94.0	95.2	148	71	208	8.6	110.0	110.5

Table 4.94. Biological peculiarities of the growth and development of ambari-hemp and jute after presowing irradiation of seeds by neutron flux in the reactor with the density of neutron flux at 1.6×10^{10} neutrons/(cm$^2\cdot$sec)

Duration of irradiation, min	Energy of germination, %	Germination, %	Density of plant stand, thousand/ha		Morphology		Yield per 10 m^2 in comparison with the control, %	
			after seedling emergence	before harvesting	plant height before harvesting, cm	diameter of stem, cm	stems	fiber
Ambari-hemp								
Control	77.0	88.7	33	31	256	9.5	100.0	100.0
5	87.0	88.7	35	37	267	9.5	115.9	110.6
15	83.0	90.5	34	35	269	9.5	105.4	102.5
55	80.7	88.5	36	37	260	9.4	109.4	102.5
Jute								
Control	92.0	94.2	142	76	207	7.8	100	100
10	93.0	94.0	153	92	213	8.2	134	132
50	96.7	96.7	155	82	212	7.9	116	122
55	95.7	96.0	159	91	211	8.5	122	115

Table 4.95. Consolidated data on the presowing irradiation of seeds of ambari-hemp, jute and dogbane

Place of research	Year, crop	Researcher	Dose of irradiation, c	Increase of yield, %	Enhancement of development
Institute of Nuclear Physics of Academy of Sciences of UzSSR and All-Union Research Institute of Cotton Cultivation	1959–1960 Ambari-hemp	S.A. Anastasov, U.A. Arifov	15–35	12.8–46.2	Additional branching
	Jute		1–7	Fiber 10–12	-do-
Institute of Nuclear Physics of Academy of Sciences of UzSSR and All-Union Research Institute of Cotton Cultivation	1927 Dogbane	Zaurov	1	Increase of yield	Increase of tillering
Institute of Nuclear Physics of Academy of Sciences of UzSSR	1959–1960 Ambari-hemp	S.A. Anastasov, U.A. Arifov	1.6×10^{10} thermal neutrons/ $(cm^2 \cdot sec)$, (5, 15, 55 min)	2.5–10.0 stems 9.5–15.9 fiber 15–32 stems 22–34	—

economically more significant than for other crops. Even an insignificant enhancement in the development of cotton would increase the pickings of prefrost raw cotton since the quality of cotton after first frost is reduced significantly. In the USSR, work on the irradiation of cotton seeds is being conducted in the Uzbek and Azerbaijan SSR.

Since 1959, at the Institute of Genetics and Plant Physiology of the Academy of Sciences of UzSSR [108–111] experiments were conducted on the cotton varieties 108F and 2525 (variety 108F is a commercial variety in almost all the crop areas of the central Asian republics). The source of irradiation was ^{60}Co. Doses of 0.5, 1, 2, 3, 4, 5, 6, 10, 15, 20 and 25 c were tried on air dried seeds with 7 to 8% moisture content. As a result of two years of trials, it has been determined as to which doses stimulate the growth and development of cotton. In addition the effects of the gamma-rays on the economic and technological indices of raw cotton fiber were also studied.

The investigations showed that positive effects on the growth and development of the cotton variety 108F were obtained through doses within a range of 0.5–2 c, which increase field germination, the actual plant stand density and raw cotton yield (Table 4.96). With further increases in the irradiation dose, the growth in plant height was intensified, but also disturbances in the natural development of the vegetative and generative organs of the plants began to emerge, leading to partial plant sterility, reduced yields and delays in ripening. Distinct adverse changes in the morphology of the cotton plant appeared as a result of irradiating seeds of the late cotton variety 2525 at 3 c and seeds of the early variety 108F with the much higher dose of 6 c.

Table 4.96. Effects of the presowing irradiation of air dried seeds by ^{60}Co gamma-rays on the growth and development of the cotton variety 108F

Dose of irradiation, c	Field germination of seeds, %	Density of plant stand, thousand/ha	Deviation from the control, thousand/ha	Yield	
				Total, q/ha	Deviation from the control, q/ha
Control	57	66.9	—	33.8	—
0.5	60	79.5	+12.6	38.1	+4.3
1.0	60	74.6	+ 7.7	35.6	+1.8
2.0	63	74.3	+ 7.4	36.9	+3.1
3.0	69	72.8	+ 5.9	33.5	−0.3
6.0	62	72.1	+ 5.2	29.7	−4.1
10.0	58	63.9	− 3.0	24.0	−9.8

The large size of the capsules (bolls) and absolute mass of seeds almost did not change through irradiation with 0.5–3 c. With much higher doses, the capsules (bolls) became smaller. The oil content of cotton seeds increased through irradiation up to 10 c. The maximum oil content in seed kernels of the variety 108F was noticed at 3 c, when the oil content of the seeds increased by 2.6% in comparison with the control (Table 4.97). Irradiating air dried seeds of cotton by ^{60}Co gamma-rays with 0.5–3 c did not deteriorate the basic technological indices of fine fibers of the late variety 2525 or the medium variety 108F. The fiber length of the variety 108F, in most cases, increased by 0.1–1 mm and by 0.6–1.4 mm in the variety 2525. Here, the fiber strength also increased.

Fertility was studied in first generation plants. The number of sterile and semi-sterile plants changed, depending on the irradiation dose.

In order to reveal the regularities of the preservation of the irradiation effects in subsequent generations, Sh.I. Ibragimov and R.I. Koval'chuk studied the second generation grown from seeds which were irradiated in different doses. The stimulation effect in the second generation increased the raw cotton of the variety 108F by 2.6–4.6 q/ha.

In 1961, commercial sowing of irradiated cotton seeds was done on 5 ha at the Akkavakskaya Experimental Station of the All-Union Research Institute of Cotton Cultivation. From unirradiated seeds, 34.9 quintals of raw cotton were obtained per hectare, whereas the yield per hectare was 36.6 quintals from the irradiated seeds, resulting in 1.7 quintals more raw cotton per hectare.

With the large-scale commercial sowing of irradiated seeds in the Fergan valley in 1962, the yield was increased by 3.7 q/ha.

At the Ya.M. Sverdlov collective farm in the Yangiyul'skii region, three 1 ha plots were sown with unirradiated and irradiated seeds (in doses of 1 and 2 c). A yield of 30.2 q/ha was obtained from the first plot, 31.8 q/ha from the second plot and 31.6 q/ha from the third. According to data from three years of experiments, the most reliable increase in the yield of raw cotton was 2–2.5 q/ha.

N.N. Nazirov and Van E-tszya worked with the medium maturing cotton varieties S-1622, 152F and 108F and an early variety 1306-DV. Seeds were sown on the day after irradiation by ^{60}Co gamma-rays in doses of 0.5, 1, 2, 5, 7, 11.5 and 15 c. The field experiments showed that irradiating air dried seeds in relatively low doses considerably accelerates the maturity of the capsules (bolls), thereby increasing the raw cotton yield of the first picking (before frosts). Moreover, the nature of the stimulation and the dose causing it were not the same for different varieties. The early variety 1306-DV gave the highest yield increase at 2 c, whereas for the medium variety S-1622—0.5 c. The yield increase and enhanced development of the late variety were considerably less. Further increases in the irradiation dose had an adverse

Table 4.97. Effects of the presowing irradiation of air dried cotton seeds by ^{60}Co gamma-rays on the quality of yield and technological indices of fiber (data 1961–1962)

Dose of irradiation, c	Mass of one capsule (boll) of raw cotton, g	Mass of 1,000 seeds, g	Oil content in the kernel of seeds, %	Technological indices of fiber			
				Output, %	Length, mm	Metric number	Strength, g
Variety 108F							
Control	7.0	134	36.6	35.4	32.3	5075	5.1
0.5	7.2	131	38.0	35.5	33.0	4930	5.3
1.0	7.0	131	37.4	35.9	33.3	5120	5.2
2.0	7.0	133	37.4	35.5	32.7	4930	5.3
3.0	6.9	132	37.8	35.3	32.3	4910	5.3
6.0	6.6	137	39.2	35.4	32.4	—	—
10.0	6.4	137	36.9	35.1	32.4	—	—
Variety 2525							
Control	3.3	—	—	28.3	43.2	—	—
0.5	3.8	—	—	28.9	43.8	—	—
1.0	3.9	—	—	29.3	44.6	—	—
2.0	3.5	—	—	29.1	44.0	—	—
3.0	3.3	—	—	28.2	23.2	—	—
6.0	3.0	—	—	28.1	42.2	—	—
10.0	3.0	—	—	28.0	49.8	—	—

effect on the maturity rates, capsule (boll) size and yield of cotton. Only in the last picking of fibers of cotton variety S-1622 was there a sharp increase as a result of an increased irradiation dose (Table 4.98).

It should also be mentioned that the early variety 1306-DV gave an almost certain increase in yield, even with irradiation at 0.5 c. Regarding other indices, this variety was found to be two to three times more resistant to the effect of gamma-rays than the medium varieties. For example, in the medium variety S-1622, deformed plants were already observed with a dose of 2 c, but sterile plants began to appear at 5 c; in the case of the early variety 1306-DV, deformed plants began at 7 c, but sterile plants appeared only at 11.5 c. Thus, there is a well-expressed, direct correlation between radioresistance and the early maturity of cotton.

A very interesting phenomenon was described by A. Uzumbaev and R.I. Koval'chuk, who irradiated air dried cotton seeds with 26 pairs of chromosomes (varieties 7059 and 2929) and 52 pairs of chromosomes with doses of 5, 10, 15, 20, 25, 30, 35 and 40 c. The irradiated seeds were soaked simultaneously with the control and then they were sown. All doses delayed the beginning of flowering and maturity: the development of the plants from the irradiated seeds was delayed by 14 days in comparison with the control. Measurements of the height of irradiated plants at the time of flowering showed a sharp increase. For example, when variety 7059 was irradiated at 20 c, the average height of the plants exceeded the control by 15 cm. If we examine individual plants grown from seeds irradiated with 25 and 30 c, we find plants whose height was 1.5–2 times more than that of the control. For example, in the variety 2929, as a result of seeds irradiated at 30 c, some plants were 212 cm in height, while the average height of the control plants was 93.3 cm. With an increased dose, the number of heterotic plants increased and in most cases they were sterile. The incidence of a huge increase in the cotton growth may be considered an effect of relative stimulation. This secondary phenomenon is linked with the fact that high doses damage the reproductive capability of plants. The extraordinary plant growth was the result of sterility caused by the high dose of irradiation. This example can serve as a confirmation of the theoretical presumption that the reproductive capability of living organisms is damaged by much lower doses of irradiation than their growth. Similar phenomenon had been described by M.M. Meisel for bacteria.

I.I. Yakobson [257] irradiated cotton seeds of the variety 8517 for 5, 10, 15 and 32 minutes after soaking them for 26 hours. Irradiation was done by a gamma source equivalent to 25 mg-eqv Ra with a platinum filter. The unit of measurement was milligram-equivalent Ra·hour. For each of the four treatments under study, doses of the following values were obtained: 2.083, 4.166, 6.249 and 13.333 mg-eqv Ra·hour. While observing the growth and development of seedlings from these irradiated cotton seeds, I.I. Yakobson

Table 4.98. Effects of gamma-irradiation of air dried cotton seeds on its yield, kg

Dose of irradiation, c	S-1622				1306-DV			
	Picking of the yield							
	first and second, 10 and 24 October	third, 10 November	last, 24 November	Total	first and second, 10 and 24 October	third, 10 November	last, 24 November	Total
Control	15.65	9.79	6.32	31.76	10.32	6.53	4.95	21.85*[21.80]
0.5	18.50	6.66	6.29	31.45	15.73	4.60	2.49	22.82
1.0	13.07	7.59	7.71	29.37* [28.37]	15.96	5.92	2.30	24.18
2.0	11.77	8.93	7.97	28.67	19.09	4.31	2.42	25.82
6.0	11.29	6.89	7.12	25.30	15.85	4.57	3.05	23.47
7.0	9.72	4.04	6.89	21.55* [20.65]	10.57	5.70	5.38	21.65
11.0	8.80	4.44	6.88	20.12	6.53	3.20	3.82	13.55
15.0	4.90	3.71	9.28	17.18* [17.89]	5.06	3.55	4.54	13.15

*Totalling errors in the Russian original. Correct totals given in brackets—General Editor.

Table 4.99. Technological characteristics of fiber from plants grown from irradiated cotton seeds

Dose of irradiation, mg-eqv Ra·hour	Width of the strip, mm	Degree of maturity of the fiber, cond. units	Area, μm^2	Strength, g
Control	21.15	2.18	110.3	4.9
2.083	22.00	2.06	108.6	4.6
4.166	23.9	1.58	102.1	3.2
6.249	24.28	1.56	105.4	3.0
13.333	24.12	1.30	90.9	2.7

noticed a definite delay in the beginning of their vegetative growth from June 16 to July 2, but then, from August 28 to September 24, these plants sharply outpaced the control in their growth. The most effective dose of irradiation was found to be 6.249 mg-eqv Ra·hour. The average mass of raw cotton in the capsule (boll) with different doses of irradiation was as below:

Dose of irradiation, mg-eqv Ra·hour	Average mass of one capsule, g
Control	4.89
2.083	4.73
4.166	7.11
6.249	7.43
13.333	6.71

The author also studied technological properties of fibers of plants grown from irradiated seeds (Table 4.100).

As may be seen in Table 4.99, the degree of maturity and strength of the fibers decrease and, hence, their technological qualities reduce. Many years have passed since this experiment was conducted. Vast material has accumulated showing the influence of prolonged after-effects of presowing irradiation of seeds, but the mechanism of this phenomenon is not yet finally resolved [258].

In Azerbaijan, R.E. Eiyubov studied the effect of gamma-rays on air dried cotton seeds of the variety 108F with doses of 1, 2.5, 7.5, 10, 15, 20 25, 30, 35, 40, 45, 50, 60 and 80 c and that of the x-rays in doses of 1, 2.5, 5 and 7.5 c. With irradiation by gamma-rays at 1–10 c, the seedlings emerged two to three days earlier in comparison with the control. Much higher doses delayed the seedling emergence by one to two days, but favorably affected the growth and development of plants. The dose of 50 c delayed fruit bearing and ripening.

D.T. Kabulov and others [115] irradiated cotton seeds of the variety 108F and hybrid No. 21 with doses of 0.2, 0.4, 0.6, 0.8, and 1.4 c. As a

Table 4.100. Consolidated data on the presowing irradiation of cotton seeds

Place of research	Year	Researcher	Dose of irradiation, c	Increase of yield in comparison with the control, %	Technological indices of fiber and seeds	Enhancement of development
1	2	3	4	5	6	7
Institute of Genetics and Plant Physiology of the Academy of Sciences of UzSSR	1958	A.M. Yakubov, Kh.U. Usmanov	2–15	—	Increase of oil content in seeds by 3%	Enhancement of development
Institute of Genetics and Plant Physiology of the Academy of Sciences of UzSSR	1959–1962	Sh.I. Ibragimov, F.P. Paiziev, R.I. Koval'chuk	0.5–2	by 4.3 q/ha	Increase of oil content in seeds by 2.6%, improvement of fiber quality	—
	1960–1961		0.5–1	by 2.6–4.6 q/ha	—	Picking of harvest before frosts, 79%
Ya.M. Sverdlov collective farm of Yangiyul'skii region	1961		1–2	by 1.6 q/ha	—	Enhancement of ripening
Institute of Genetics and Plant Physiology of the Academy of Sciences of UzSSR	1962	N.N. Nazirov, Van E-Tszya	1–2	by 1.6–4 q/ha	—	Enhancement of development

(*Contd.*)

Table 4.100 (*Contd.*)

1	2	3	4	5	6	7
Institute of Soil Science and Agro-Chemistry of the Academy of Sciences of AzSSR	1958 1959 1960 1959 1960	R.E. Eiyubov	2.5–10 1–10 1–10 1.0– 7.5	by 4.4 q/ha by 4.8 q/ha by 2.35 q/ha by 1.7–2 q/ha	— — — —	Enhancement of ripening
Botanic Garden of Academy of Sciences of Tadjik SSR	1961	N.N. Bazhenova	0.5-1	Increase of root length by 18%, height of the seedlings by 15%	—	—
Collective Farm Kommunist of Post-Dargomskii region	1960	D.G. Kabulov, M.M. Muminov, A.K. Aleksanyan	0.8	by 8.42 q/ha	—	—
Institute of Soil Science of Academy of Sciences of UzSSR	1961	D.G. Kabulov, M.M. Muminov, A.K. Aleksanyan	0.4–0.8	Variety 108F by 44–49%; hybrid No. 21 by 39–51%	—	—
Institute of Soil Science of the Academy of Sciences of UzSSR	1961	D.G. Kabulov, M.M. Muminov, A.K. Aleksanyan	Action by thermal neutrons	by 3–13%	—	—

result the following conclusions were made: presowing irradiation facilitates cotton growth; the quantity of capsule-elements in the variety 108F increased by 8.6%, while the number of capsules (bolls) that remained on the plants to maturity increased almost twofold in comparison with the control; the average mass of capsule (boll) was 7.8–8.8 g, as compared to 7.5 g in the control; hybrid No. 21 increased more than variety 108F in all indices.

Generalizing the results of investigations on the presowing irradiation of cotton conducted independently of each other by different researchers (Table 4.100), we see that they all reported comparatively uniform changes over a five-year period. These were manifest in the increased yield of raw cotton and fiber, as a result of irradiation in doses to 3 c, because of the formation of a large number of capsules (bolls). According to the data from R.E. Elyubov, the optimal stimulation doses were somewhat higher—up to 5 c. All researchers except N.N. Yakobson noticed an improvement in the technological indices of the fibers and enhanced development of plants, facilitating an early harvest of more capsules (bolls) before frost. The data obtained demonstrate the need to complete the research on presowing irradiation to use this measure in practical cotton growing.

OILSEEDS AND VOLATILE OILSEED CROPS

The presowing irradiation of oilseed crops, except sunflower, was done only on an insignificant scale, while we have only a few experiments reported on volatile and essential oilseed crops.

Sunflower. In the work of Shull and Mitchell [350], we find the first reference to the results of presowing irradiation on sunflower seeds. The seeds were subjected to x-rays, using an aluminum filter, in a dose with a strength of about 38 r/m and kept 30 cm from the source. Before irradiation, the seeds were kept in a chamber on a layer of moist cotton soaked in distilled water and at 22°C for 24 hours. Then the pericarp was removed from the seeds and they were irradiated for 1–10 minutes. In the irradiation treatments for 1–3 minutes, the significant stimulation of plant growth and development corresponded to the irradiation dose of 38–114 r. In the dose of 380 r, which the seeds got for ten minutes, there was a sharp suppression in the plant development associated with different deviations from the normal development of morphological characters, color, change in the shape of the leaf blade, fasciation of stems and leaves and so on.

Experiments on irradiating soaked sunflower seeds were conducted also by Wetterer [363], who irradiated seeds at 5, 10, 20 and 40 NED. With all irradiation doses, the plants showed retarded growth in comparison with the control. It may be presumed that Wetterer's conclusion, on the damaging effects of irradiation, in all doses he tried, was based on a wrong selection of doses, which did not provide him with any stimulation effect for this crop.

At the Institute of Biophysics of the USSR Academy of Sciences, T.E. Guseva [83] conducted experiments for three years on the presowing irradiation of the seeds of sunflower variety Peredovik. The effects of doses at 1, 4, 16 and 40 c were tested under laboratory conditions. The presence of the stimulation effect of gamma-rays was established in a range from 1–4 c, in which the increased root length of the sprouts exceeded the control by 12 and 16% on an average (Table 4.101). The lethal dose for sunflower was found to be 200 c.

Measurements of the aerial mass of plants grown in pots, as an average of three experiments on seeds irradiated at 4 c, show an increase of 13% in comparison with the control (Table 4.102).

Table 4.101. Length of the roots of sunflower plants on the 9th–10th day after sowing, depending on the irradiation dose

Dose of irradiation, c	Experiment 1		Experiment 2		Experiment 3	
	mm	%	mm	%	mm	%
Control	40±0	100	210±6.10	100	159±5.8	100
1	60±2.9	150	217±1.8	103	166±10.4	105
4	51±1.7	128	229±3.9	109	176±4.9	111
16	28±2.0	98	195±4.1	93	154±4.4	97
40	19±0.5	47	176±6.5	94	101±5.3	64

Dose of irradiation, c	Experiment 4		Experiment 5		Average	
	mm	%	mm	%	mm	%
Control	132±7.2	100	84±4.1	100	125	100
1	181±14.3	137	80±2.3	95	141	112
4	166±3.3	126	102±2.7	121	145	116
16	148±9.7	113	54±2.5	64	116	93
40	95±9.4	73	30±4.0	36	84	67

Table 4.102. The mass of the aerial part of sunflower plants depending on the dose of irradiation

Dose of irradiation, c	Experiment 1		Experiment 2		Experiment 3		Average	
	g	%	g	%	g	%	g	%
Control	7	100	7.2	100	8.7	100	7.6	100
1	7	100	7.8	108	8.1	93	7.6	100
4	7.9	113	8.6	118	9.4	107	8.6	113
16	6.8	97	7.2	100	8.9	101	7.6	100
40	5.0	72	5.4	75	7.9	90	6.1	80

Results of the pot culture experiments show that for sunflower seeds irradiated with 1–4 c, there was a 21–25% increase in the yield, in comparison with the control, and a reduction in the weight of 1,000 seeds (Table 4.103).

In field experiments the irradiation of sunflower seeds was done over a three-year period at the experimental base of the All-Union Research Institute of Oilseed Crops. Only the dose of 4 c permitted as 6–7% increase in seed yield in 1966 and 1968. In 1967, there was no yield increase. The presowing irradiation of air dried seeds at 16 and 40 c, led to the development of deformed capitulum (Fig. 4.10), whose quantity was 15% at 16 c and 96% at 40 c. It was interesting to note that morphological changes appeared in the formation of additional capitulum as a development of multicob forms in corn, the formation of capsule (boll) bunches in cotton and the emergence of additional branches in other crops.

Table 4.103. Yield of sunflower seeds by growing them in pots, depending on the dose of irradiation

Dose of irradiation, c	Yield of seeds		Weight of 1,000 seeds	
	g	%	g	%
Control	16.6±0.8	100	64.4	100
1	20.1±1.7	121	55.8	88
4	20.8±0.9	125	55.9	87
16	16.8±1.8	96	54.9	85
40	12.3±1.2	74	62.9	9.8

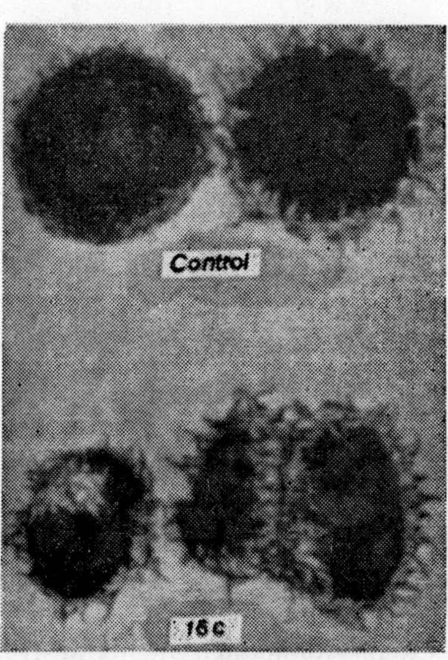

Fig. 4.10. Change in the shape of capitulum of sunflower after seed irradiation with 16 c.

The proteins, amino acids, oil and husk contents of seeds and the weight of 1,000 seeds were determined from the yield. With irradiation at the stimulation doses (1–4 c) the mass of 1,000 seeds comprised 98–101% of the control, but with the suppressing doses of 16 and 40 c, it increased by 10–27%. With a much higher absolute mass of seeds, their husk content increased to 1.7%. Increased husk contents of seeds is a negative factor. With the suppressing dose, the oil content of seeds reduced considerably—by 1.6–2.8%, but there was no change in this index with the stimulation dose. Results of the determination of protein and amino acid contents in the seeds are given in Table 4.104.

As may be seen from the data presented in the above table, at 16 c, the total content of proteins increased by 16% and amino acids by 23%. Simultaneously with the increase of protein and amino acids, there was a reduction in oil content and seed yield. Thus, a direct, widely prevalent correlation was observed between these processes. Increased protein content may be explained by the fact that during the oil formation process the proteins are not completely used by irradiated plants.

Table 4.104. Content of proteins and amino acids in sunflower seeds depending on the dose of irradiation

Dose of irradiation, c	Total proteins			Amino acids	
	%	In percentage of the control	Deviation from the control	Cond. unit	In percentage of the control
Control	19.21	100	0	0.044	100
1	19.72	104	+0.51	—	—
4	19.41	102	+0.20	0.035	80
16	21.94	116	+2.73	0.054	123

The presowing irradiation of sunflower seeds was also done for a four-year period by the Research Institute of Agriculture of the AzSSR [201] under conditions of the Apsheronskoe Peninsula. The seeds of sunflower variety Zhdanovskii were irradiated. The laboratory as well as the subsequent field experiments showed the possibility of increasing the sunflower seed yield by 10–18% and a much higher oil content. The established stability of the appearance of the stimulation effect shows the possibility of large-scale commercial tests on the presowing irradiation of seeds of this crop under conditions of AzSSR.

Two years of large-scale commercial sowing of irradiated sunflower seeds of the variety Kustanaiskii 91 in the Pavlodar province of the KazSSR on

1,000 hectares made it possible to increase the seed yield by 8.3 q/ha, comprising about a 20.3% increase over the control. In 1972, when commercial sowing of the presowing irradiated seeds was repeated on 940 ha, there was an increase of 14% in the yield [158].

In the MSSR [47], experiments were conducted on the presowing irradiation of the sunflower variety Peredovik. The optimal stimulation dose was 2 c, which resulted in accelerated plant development by two to four days in comparison with the control and increases in the anthodium diameter by 14.4% and seed yield by 18.2% (4.5 q/ha).

In the experimental-cum-demonstration farm of the Umanskii Agricultural Institute, the presowing irradiation of sunflower seeds of the variety Peredovik was done with doses of 0.3 and 2 c on the GUT-400 unit with ^{60}Co gamma-rays at a strength of 700 r/min. Sunflower ripening was earlier by four and two days with irradiation doses of 0.3 and 2 c, respectively. As a result of irradiation, the diameter of the anthodia increased by 2.17 cm which comprised 10.6% in comparison with the control. The yield with the dose of 0.3 c was 16.8 q/ha, i.e., it was 2.6 q/ha or 18.3% higher than that of the control. With the dose of 2 c, the yield, although exceeding the control, was somewhat less than with the other doses. After irradiation at 0.3 c, the oil content of the seeds increased by 0.3%, but this increase was not stable.

Mustard. It is one of the most radioresistant oil seed crops. For this crop, the lethal dose is 750 c, under laboratory conditions and it has a very wide range of stimulation doses—1–40 c. With the irradiation of seeds at 300 c, there is some suppression of growth and development of plants: flowering started six to eight days later, which subsequently adversely affected the entire life-cycle of the plants. The experimental data on yield estimates of mustards are shown in Table 4.105.

As evident from the above data, irradiation of seeds at 1–40 c led to a big seed yield increase per plant—up to 34–41% in comparison with the control. With 300 c, there was no yield increase, but the number of pods increased by 101%. However, the pods formed in this treatment had underdeveloped seeds which were mostly sterile. An increase in oil content by 0.4–4.6% was observed through irradiation of mustard seeds at the stimulation dose.

The seeds collected from irradiated plants were sown again to study the after-effects (Table 4.106). In treatments where seeds in the M1-generation were irradiated at the stimulation dose (1–40 c), the seed yield in the M2-generation exceeded the control by 13–21%, however, the yield increase here was less than the increase obtained from the M1-generation. Irradiation with 300 c reduced the yield by 8%.

Individual selections were made from the progeny of seeds irradiated with 300 c, of which three plants had very large seeds while the other three were high yielding plants. In 1968, an M3-generation was obtained

Table 4.105. Yield of mustard depending on the dose of irradiation

Dose of irradiation, c	No. of plants harvested	No. of pods per plant		Mass of seeds per plant	
		number	%	g	%
Control	74	13.1	100	0.29±0.03	100
1	67	13.6	104	0.39±0.04	134
4	69	16.1	123	0.41±0.01	141
16	70	14.6	111	0.38±0.03	131
40	74	16.2	124	0.41±0.03	141
100	60	16.1	123	0.30±0.03	104
300	23	26.3	201	0.30±0.03	104

Table 4.106. Effects of irradiation on the yield of mustard seeds in the M2-generation

Dose of irradiation, c	Mass of seeds on one plant	
	g	%
Control	0.24±0.01	100
1	0.27±0.01	113
4	0.29±0.03	121
16	0.28±0.03	117
40	0.27±0.01	113
100	0.24±0.01	100
300	0.22±0.02	92

from seeds that were irradiated with 300 c. The selections of individual plants made on the basis of yield and the large size of seeds exceeded the control with respect to the mass of seeds from one plant (by 17–21%) and the total mass of the seeds [83].

Castor. The presowing irradiation of seeds of this crop was first done at the Institute of Biophysics of the USSR Academy of Sciences by T.E. Guseva under laboratory conditions, in pots and in field experiments. Effects from doses of 1, 4, 16 and 40 c, were examined and the doses of 1 and 4 c were found to be optimal. They were established as a result of five laboratory experiments (Table 4.107). As is evident from the table, the roots increased from these doses, on the average, by 12 and 14% in comparison with the control, while the increase in plant height was 4–18% (Table 4.108).

A dose of 300 c was found to be lethal for this crop.

In 1966 and 1967, the results of laboratory and pot culture experiments were verified under field conditions at the experimental base of the All-Union Research Institute of Oilseed Crops in Krasnodar (Table 4.109). As is evident from the data presented in the table, very insignificant and unreliable yield increases were obtained with irradiation at the stimulation doses of 1 and 4 c.

Table 4.107. Root length of castor on the 27–30th day after irradiation of air dried seeds

Dose of irradiation, c	Experiment 1		Experiment 2		Experiment 3		Experiment 4		Experiment 5		Average	
	mm	%	mm	%	mm	%	mm	%	mm	%	mm	%
Control	160±3.0	100	248±4.9	100	298±9.0	100	179±4.1	100	202±3.6	100	217	100
1	231±5.0	144	289±6.6	117	313±7.0	105	204±4.7	114	197±7.7	97	247	114
4	216±4.0	135	288±1.2	116	317±6.0	106	200±3.2	112	218±6.4	108	244	112
16	177±0.8	110	247±3.5	100	277±6.0	93	193±4.5	108	232±8.9	115	225	104
40	100±6.6	63	—	—	—	—	162±4.6	90	173±2.4	86	145	66

Table 4.108. Stem height of castor grown from irradiated seeds in pots on 25–26th day

Dose of irradiation, c	Experiment 1		Experiment 2		Experiment 3		Average	
	mm	%	mm	%	mm	%	mm	%
Control	288±3.4	100	187±4.1	100	166±1.7	100	194	100
1	211±16.4	97	194±3.7	104	167±0.9	100	191	98
4	276±12.0	121	208 ±4.7	112	204±5.5	123	229	118
16	248±10.4	109	208 ±4.4	112	176±4.3	106	211	109
40	247±11.9	109	190±8.8	102	169±3.7	102	202	104

Table 4.109. Seed yield of castor depending on the dose of irradiation

Dose of irradiation, c	1966			1967			Average		
	q/ha	%	Deviation from the control	q/ha	%	Deviation from the control	q/ha	%	Deviation from the control
Control	13.7	100	0	12.4	100	0	13.1	100	0
1	14.2	104	+0.5	—	—	—	—	—	—
4	14.2	104	+0.5	12.6	102	+0.2	13.4	102	+0.3
16	13.4	98	−0.3	12.1	98	−0.3	12.8	98	−0.3
40	—	—	—	11.2	90	−1.2	—	—	—
80	—	—	—	4.6	37	−7.8	—	—	—

Analysis of the yield, however, showed a unique reaction of castor to irradiation. With cultivation in the USSR, this subtropical plant produces mature seeds only on the central raceme. Because of the effects of irradiation, there is intensive additional branching of the first order and the number of branches increases by 24% in comparison with the control. Branches of the second and third order, which were absent in the control, appeared simultaneously (Fig. 4.11). Each of these additional branches produced a flowering raceme, which did not mature or produce a yield under the conditions in Krasnodar. These unproductive racemes weaken the development of the central raceme and reduce the yield of fully developed seeds. However, the weight of green mass increased sharply by virtue of the profuse additional branching, resulting in an increase up to 60% over the control. This phenomenon, in which the stimulation of some parts of the plants without any practical importance occur, is termed by us as relative stimulation. Nevertheless, the productivity estimate of such plants is of theoretical interest because it unveils different aspects of the stimulation effects of irradiation [83].

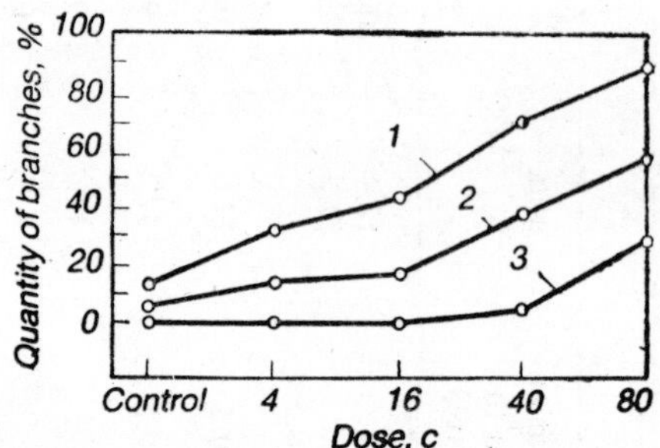

Fig. 4.11. Change of branching in castor plants depending on the dose of irradiation: *1*—branches of the first order; *2*—branches of the second order; *3*—branches of the third order.

Soybean. Research on the presowing irradiation of soybean was conducted by several authors, however, they were not extended to field experiments with concluding yield estimates. Long and Karsten [317, 318] irradiated soybean seeds and got an increased green mass. M.M. Tushnyakova and M.A. Vasilevskii [233] pointed out that the presowing irradiation of soybean seeds resulted in enhanced soybean maturity by 12 days, however, there is no mention of yield data in this work.

At the Institute of Biophysics of the USSR Academy of Sciences, laboratory experiments on the presowing irradiation of the seeds of soybean variety Komsomolka were conducted by T.E. Guseva [83]. The seeds were irradiated by ^{60}Co gamma-rays in a wide range of doses: 0.25, 0.510, 1, 2, 4, 10, 40 c. The stimulation effect was observed at 0.25–2 c. The dose of 40 c was found to be lethal (Table 4.110).

When irradiated seeds were grown in pots and in the field, the stimulation effect was found in the much lower doses of 0.25–1 c.

Table 4.110. Root length of soybeans with the gamma-irradiation of seeds

Dose of irradiation, c	Experiment 1		Experiment 2		Average	
	mm	%	mm	%	mm	%
Control	116±3.8	100	131±1.6	100	124	100
0.25	142±4.1	122	142±1.9	109	143	116
0.50	133±2.3	115	140±2.1	107	137	111
1.0	157±2.8	136	149±1.4	114	153	123
2.0	133±2.6	115	143±1.1	109	138	115
4.0	118±2.7	102	131±3.0	100	125	102
10.0	107±4.1	92	107±2.2	82	107	101
40.0	19±2.4	10	22±1.3	17	21	12

In Hungary [354], seeds of the soybean variety Chippeva, with a moisture content of 8.9%, were irradiated in a wide range of doses—from 0.25 to 20 c, with the strength of 40 r/min; the irradiated seeds were sown 24 hours later. The purpose of the experiment was to reveal certain biochemical changes in soybeans due to the effects of irradiation. The protein contents of green plants, the proteins and oil in mature seeds and the content of microelements Zn, Fe, Mn and Cu were determined (Tables 4.111 and 4.112).

The data of Table 4.112 show the increased protein content in green plants and in mature seeds with high irradiation doses to seeds (15–20 c) and the insignificant increase of these substances in the green mass through irradiation at the stimulation dose (0.025–0.1 c). As may be seen from Table 4.113, the increase in the content of microelements as a result of presowing gamma-irradiation was observed with a wide range of doses and the highest increase of Zn and Mn content occurred with a low stimulation dose (0.025 c).

Changes in the content of proteins and oil in soybean seeds were also noticed by Williams and Khanvai. These authors made yield estimates of soybean through irradiation in these doses. Increased yield was observed only with doses to 2 c (Table 4.113).

Of great interest are the data on the changed flowering periods of soybean grown from irradiated seeds, which shows drastically enhanced

Table 4.111. Content of proteins and oil in soybean plants grown from irradiated seeds

Dose of irradiation, c	Green plants		Green leaves		Mature seeds			
	Proteins				Proteins		Oil	
	%	In percentage of the control	%	In percentage of the control	%	In percentage of the control	%	In percentage of the control
Control	28.15	100	14.25	100	22.15	100	23.66	100
0.025	28.28	100	14.87	104	22.26	100	23.83	101
0.1	28.46	101	15.10	106	21.99	99	23.98	101
0.5	28.56	101	14.19	99	21.56	97	24.15	102
1.0	28.05	99	14.96	105	21.65	98	24.30	103
2.0	28.42	100	14.85	104	22.42	101	23.92	101
5.0	28.67	101	17.79	104	22.80	103	23.63	100
10.0	28.79	102	14.52	102	23.30	105	23.87	101
15.0	30.99	110	18.11	127	24.34	110	24.57	104
20.0	33.84	120	21.58	151	26.43	119	23.69	100

Table 4.112. Content of microelements in soybeans through the presowing irradiation (as per dry mass)

Dose of irradiation, c	Zn		Fe		Mn		Cu	
	g	%	g	%	g	%	g	%
Control	71	100	205	100	39	100	11	100
0.025	100	141	125	119	51	131	13	118
0.1	84	118	114	108	43	110	12	109
0.5	79	111	111	106	43	110	11	100
1.0	64	90	108	103	41	105	11	100
2.0	79	111	122	116	46	118	12	109
5.0	79	111	120	114	51	131	13	118
10.0	84	118	124	118	47	120	13	118
15.0	67	94	111	106	36	92	11	100
20.0	62	87	106	101	34	87	9	82

Table 4.113. Soybean yield obtained as a result of the presowing irradiation of seeds

Dose of irradiation, c	Average grain yield, g/pot	In percentage of the control
Control	59.73 ± 1.49	100
0.025	60.70 ± 0.56	101
0.1	63.01 ± 0.64	105
0.5	61.12 ± 0.54	102
1.0	63.40 ± 0.69	106
2.0	61.07 ± 0.63	102
5.0	57.52 ± 0.38	96
10.0	56.33 ± 0.62	94
15.0	55.55 ± 1.22	93
20.0	37.02 ± 1.0	62

flowering by almost 14 days. These data very closely coincide with earlier data from M.M. Tushnyakova and M.A. Vaslevskii who observed enhanced flowering in soybean by 12 days (Table 4.114).

Counting the number of flowers blooming on different days over a two-week period and expressing them in percentage to the control, the authors showed that during the first few days, in the case of plants whose seeds were irradiated at the stimulation dose, the flower blooming is 500, 411 and 244% in comparison with the control.

Peanut. The determination of the optimal stimulation doses for the peanut variety Adig was done under laboratory conditions and in the greenhouse of the Institute of Biophysics of the USSR Academy of Sciences [83], through tests on a wide range of doses of ^{60}Co gamma-rays from 0.25 to 80 c.

From the data presented in Table 4.115, it is evident that the stimulation, as measured in plant height, appeared with the irradiation of seeds with 0.25–0.50 c.

Table 4.114. The number of blooming flowers of soybean grown from irradiated seeds, in comparison with the control, %

The day of estimate	Dose of irradiation, c									
	Control	0.025	0.1	0.5	1	2	5	10	15	20
1st	100	500	411	244	78	155	78	—	—	—
2nd	100	326	318	211	100	133	74	—	—	—
3rd	100	163	186	149	83	111	59	15	—	—
4th	100	142	176	141	92	126	79	33	0.8	—
5th	100	126	144	123	104	126	76	33	1.4	0.5
6th	100	101	106	106	85	90	62	28	1.9	0.3
7th	100	102	102	105	85	90	63	85	3.2	0.7
8th	100	104	105	106	88	90	66	39	4.6	0.9
9th	10C	104	107	109	92	92	73	44	6.2	0.8
10th	100	107	111	106	94	91	80	52	9.3	0.9
11th	100	106	109	107	100	98	87	60	15.0	2.7
12th	100	109	110	106	101	96	87	65	18.0	3.9
13th	100	108	108	105	100	95	86	65	48.0	4.4
14th	100	109	108	106	100	96	87	65	19.0	5.6

Table 4.115. Estimate of plant height in peanut with the presowing irradiation of seeds

Dose of irradiation, c	Experiment 1		Experiment 2		Experiment 3		Average	
	mm	%	mm	%	mm	%	mm	%
Control	66±3.2	100	72±2.7	100	97±2.8	100	78	100
0.25	69±2.8	105	80±2.3	111	110±3.1	114	86	100
0.50	76±2.6	115	78±2.0	108	103±3.5	106	86	110
1.0	54±4.0	82	68±2.4	95	91±4.2	94	71	91
2.0	48±3.7	72	60±3.1	84	76±3.6	78	61	78
4.0	42±4.3	64	46±2.9	64	70±2.8	72	53	68
8.0	25±1.2	38	23±1.8	32	44±2.0	45	31	40

Comparing the lethal and stimulation doses for the above oilseed crops, we see great differences between them—from 750 to 40 c (Table 4.116).

It is also evident from the data presented in the table that the range of stimulation doses narrows with the reduced value of the lethal dose, i.e., parallel to a reduction in radiosensitivity.

Table 4.116. Comparative radiosensitivity of some oilseed crops

Crop	Variety	Dose, c	
		lethal	stimulation
Mustard	Skorospelka	750	1–40
Castor	VNIIMK-165	300	1–16
Sunflower	Peredovik	200	1–4
Peanut	Adig	80	0.25–5
Soybean	Komsomolka	40	0.25–2

The wide range of stimulation doses in the presowing irradiation of seeds can be considered a positive factor because it is more difficult to establish optimal doses and to get good reproducibility of the stimulation effect for plants with a narrow range of stimulation doses. The lethal dose value under field conditions was considerably lower than in the laboratory experiments, particularly for mustard $LD_{100} = 300$ c and for sunflower $LD_{100} = 40$ c. Results of experiments on oilseed crops are given in Table 4.117.

Coriander. Laboratory experiments on this crop were conducted at the Institute of Biophysics of the USSR Academy of Sciences, to find the doses which stimulate plant growth and development. Air dried seeds of coriander were irradiated by ^{60}Co gamma-rays in doses of 0.1–40 c on the GUBE unit with the dose strength at 450 r/min. With irradiation at 0.5–0.1 c, there was a stimulation of growth and development in seedlings grown in Petri dishes. The criteria for the evaluation of the stimulation effect were measurements of the radicle length of seedlings. In plants grown from seeds irradiated at 1 c, the length of radicles exceeded the root length of control seedlings by 18%. Field experiments were conducted at the All-Union Research Institute of Oilseed Crops in Krasnodar, in which there was an 18% increase in the coriander yield with the dose of 1 c.

It has been established by the work from the Nikitskii Botanic Garden [244] that irradiation of the coriander varieties Luch and A-247 with doses of 0.5–3 c stimulates not only the growth and development of seedlings but also increases the seed yield by 25% and there is an improvement in the quality of essential volatile oils with consequent increase in it by 2–3% of one of the most valuable components, viz., linalool. However, with the

Table 4.117. Consolidated data on the presowing irradiation of the seeds of oilseed crops

Place of research	Year, crop	Researcher	Dose of irradiation, c	Increase of yield in comparison with the control, %	Biochemical changes	Enhancement of development
1	2	3	4	5	6	7
—	1963 Sunflower	Shull, Mitchell	0.03–0.11	Significant stimulation	—	—
Institute of Biophysics of the USSR Academy of Sciences	1969 Sunflower	T.E. Guseva	1–4	Length of roots 12–16	—	—
	1970 Sunflower		1-4	Seed yield 21–25	Increase in the content of total proteins and amino acids	—
Research Institute of Agriculture of AzSSR	1968–1970 Sunflower	R.R. Riza–Zade	2	10–18	Much higher oil content	—
Kishinev Agricultural Institute named after M.V. Frunze	1968–1970 Sunflower	V.G. Bordyuzhevich, V.N. Lysikov, K.I. Sukach, N.M. Berezina, D.A. Kaushanskii	2	18.2	Much higher oil content	Enhancement of development
Pavlodar Directorate of Agriculture of KazSSR	1971 Sunflower	A.V. Kalashnikov, Yu.A. Martem'yanov	2	20.3	—	—
	1972 Sunflower		2	14	—	—

(Contd.)

Table 4.117 (*Contd.*)

1	2	3	4	5	6	7
Experimental Farm of Umansk Agricultural Institute	1972 Sunflower	A.P. Ivanov, L.A. Lospan, A.T. Lisovets, I.T. Voichenko	0.3–2	16.8	Increase o oil content by 0.3%, increase unstable	Enhancement of ripening by 2–4 days
Institute of Biophysics of the USSR Academy of Sciences	1969–1971 Mustard	N.M. Berezina, T.E. Guseva, L.N. Kharchenko	1–4	34–40	Increase of oil content by 0.4–4.6%	—
Institute of Biophysics	1969–1971 Castor	N.M. Berezina, T.E. Guseva	1–4	—	Pseudo-stimulation	—
—	1936 Soybean	Long, Karsten	1	Increase of green mass	—	—
Institute of Biophysics of the USSR Academy of Sciences	1969 Soybean	T.E. Guseva	0.25–2	Length of roots 11–23	—	—
Hungary	1971 Soybean	Stan, Croitur, Keppel	—	—	Increase of protein content in leaves by 4–6%, oil in the seeds by 1–3%	Enhancement of flowering by 14 days
Institute of Biophysics of the USSR Academy of Sciences	1969 Peanut	T.E. Guseva	0.25–0.5	—	—	—

Table 4.118. Consolidated data on the presowing irradiation of seeds of essential volatile oil crops

Place of research	Year, crop	Researcher	Dose of irradiation, c	Increase of yield in comparison with the control, %	Enhancement of development	Biochemical changes
State Nikitskii Botanic Garden	1962 Lavender	S.G. Malyarenko	3	—	Increase of germination percentage of seeds which germinate with difficulty	—
Institute of Biophysics of the USSR Academy of Sciences	1961 Peppermint	N.M. Berezina	1	Green mass 21	—	—
All-Union Research Institute of Oilseed Crops	1963 Coriander	V.P. Umen	1	Seeds 18	—	—
State Nikitskii Botanic Garden	1972 Coriander	N.G. Chemarin, A.I. Arinshtein, L.A. Kichanova	0.5–3	25	—	Increase in the content of linalool by 2–3%
All-Union Research Institute of Oilseed Crops	1970 Coriander	L.A. Kichanova	1	12	—	—

stimulation dose, the total content of essential volatile oils did not increase. It increased only through a higher dose of irradiation—7–10 c. According to the data from these researchers, the lethal dose for coriander was 20 c.

Lavender. Lavender seeds, one of the most valuable essential oil bearing plants, belong to the group of seeds which germinate with great difficulty and require prolonged and cumbersome stratification, through which, however, we get a very low percentage of germinated seedlings. Lavender seeds were irradiated at the Nikitskii Botanic Garden [154] by ^{60}Co gamma-rays to accelerate their germination. Seeds of two varieties, viz., Record N-761 and Prima N-328, both stratified and unstratified, were irradiated with doses of 0.1, 0.3, 0.5, 1, 3, 5, 15, 20 and 50 c, with the strength of 500 r/min. The irradiated seeds were germinated in Petri dishes over a 30-day period.

Through the optimal stimulation irradiation dose of 0.5–3 c, the germination of stratified seeds increased to 88 and 62% respectively. In several years of earlier experiments on the study of the biology of the germination of lavender seeds, it had not been possible to get such a high percentage of germination through any treatment.

Further observations on the growth and development of seedlings showed that the length of roots, the height of the seedlings and the dry mass of plants grown from seeds irradiated at 1 c were much greater than in the control. Irradiation at 3 c stimulated only the root length and much higher doses stunted the growth and development of seedlings. Plants grown from the seeds that were irradiated at 15 c and more died after 28–30 days.

Peppermint. Rhizomes of the peppermint variety Prilukskaya-6 were irradiated at the Institute of Biophysics of the USSR Academy of Sciences by ^{60}Co gamma-rays with doses of 0.5–8 c, with strength of the dose at 750 r/min. It must be mentioned that with a large number of reports on the presowing irradiation of seeds, very little analogous work had been conducted on vegetative plant material. Irradiating peppermint rhizomes with 4 c showed that there was no germination of any buds at all. With their irradiation with 1 c, the plant growth was stimulated.

The quantity of sugars was determined in the rhizomes of control and irradiated plants. This index is very important because with a much higher sugar content, the rhizomes are more frost resistant. However, the observations showed the presence of a distinct negative correlation between the intensity of the growth of rhizomes and their sugar content: at 4 c, the buds of rhizomes did not germinate, but the sugar content in rhizomes was considerably higher than in the control; at the stimulation dose of 1 c, the sugar content in the rhizomes was lower than in the control.

Table 4.118 shows the results of experiments on essential volatile oil crops.

5. Radiation Technology for the Presowing Irradiation of Seeds

SOURCES OF IONIZING RADIATIONS

The development of the method of presowing irradiation of seeds as an agronomical measure satisfying the requirements of agricultural production is linked with the evolution, creation and application of different types of ionizing radiation sources. Earlier, x-rays were used to study the stimulation effect of radiations. Now ^{137}Cs sources of radiation are widely used in practice, but for research in the USSR and abroad ^{137}Cs and ^{60}Co are being used as sources of radiation.

Scheme 5.1 shows the sources of radiation which can be applied in different branches of industry. As the units with radiation contours and heat releasing elements must be close to the nuclear reactor, it is evident that ^{60}Co and ^{137}Cs are the only sources of radiation which can be used to create a mobile radiation technology.

The main prerequisites now required from the sources of ionizing radiation used for the presowing irradiation of seeds are as follows: the seeds must receive a definite stimulation dose distributed comparatively uniformly through the entire volume of the sowing material; undesirable reactions accompanying the appearance of induced activity in the irradiated material must be absent and the given interval of the strength of dose must be ensured. The possibility of acquiring radiation sources in the required quantity, the comparatively high penetrating power of radiation and the energy of gamma-irradiation without inducing activity in the irradiated seeds, the physical and mechanical properties of the radiation sources that would enable us to use them with ease, for a long time under agricultural production conditions—all these factors led to the extensive use of ^{137}Cs and ^{60}Co gamma sources of radiation. The production of these sources of radiation is progressively increasing in our country as well as abroad. It is adequate to mention that, according to the data available in the literature, the production of ^{137}Cs in the USA with the commissioning of the plant "Isochemical" must reach 25 m curie in a year. Table 5.1 shows the production data [337] of ^{137}Cs in France up to 1980. The growth of the production of ^{60}Co can be seen by the USA example which produced 1 m curie ^{60}Co in 1961, 2 m curie in 1962; 13 m curie in 1966, and up to 40 m curie in 1972 [215, 330, 331].

Scheme 5.1

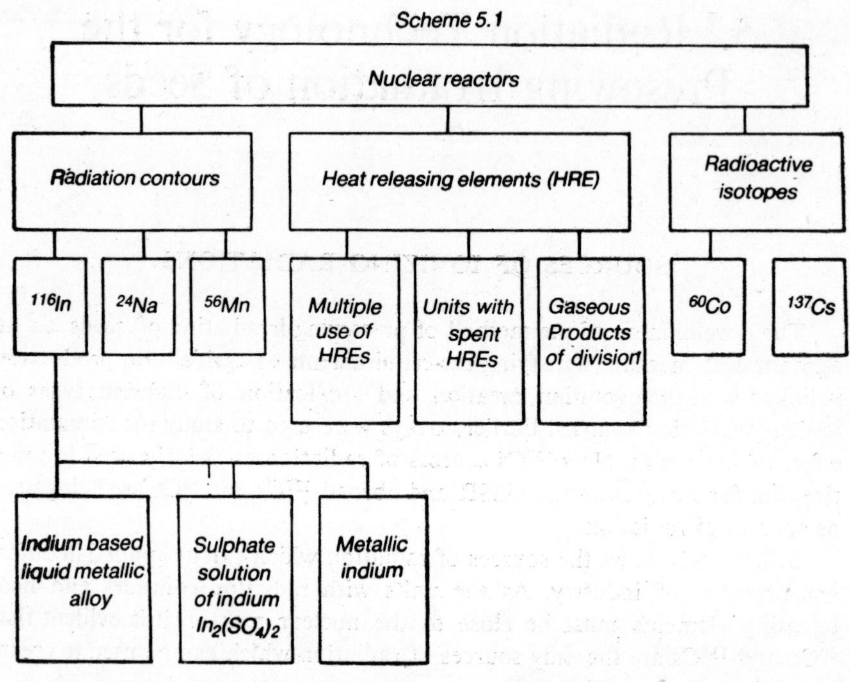

Table 5.1. Production of ^{137}Cs in France

Year	Production, m curie/year	Cost of 1 curie, francs	
		in powder	in capsule
1971	0.5	6.2	9.2
1974	2.5	1.5	2.3
1977	5.0	1.1	1.6
1980	10.0	0.8	1.1

Because mobile gamma irradiation units must have the required mobility for large-scale use under different soil and climatic conditions, ^{137}Cs and ^{60}Co radioisotopes are the main sources of ionizing radiation.

Table 5.2 shows the characteristics of radioisotopes of cobalt and cesium, which must necessarily be taken into consideration during their application.

Half life is the time during which half the initial atoms of a radioisotope disintegrate. It is a constant value for a given isotope and it is practically independent of temperature, pressure and physical state of the matter. From the aspect of using radiation sources in the units for presowing irradiation of seeds, the sources with longer half lives are preferred, because they increase the reliability of the units, reduce the number of irradiator recharges and consequently the radiation hazards, besides ensuring more stable seed treatment.

Table 5.2. Main characteristics of radioisotopes of cobalt and cesium used as sources of gamma-radiation

Isotope	Half-life period	Energy of gamma-radiation, MeV	Output of gamma-radiation upon disintegration, %	Gamma-constant r-cm²/(hour-m curie) differential	Gamma-constant r-cm²/(hour-m curie) total	Gamma-equivalent of 1 m curie of isotope mg-eqv. of radium
^{57}Co	270 days	0.700	0.2	0.08		
		1.137	11.8	0.076	0.576	0.07
		0.123	88.0	0.492		
		0.014	—	—		
		2.158	1.2×10^{-3}	—		
		1.333	100	6.82		
^{60}Co	5.63 years	1.172	99	6.11	12.93	1.54
		0.825	2.8×10^{-3}	—		
$^{137}Cs+$ ^{137}Ba	29.68 years	0.661	82.5	3.1	3.1	0.37
^{134}Cs	2.1 years	1.370	3.3	0.217		
		1.170	2.5	0.154		
		1.040	1.5	0.084		
		0.960	0.6	0.031		
		0.801	10	0.449	8.58	1.02
		0.796	80	3.568		
		0.605	95	3.296		
		0.570	0.119	0.004		
		0.569	14	0.451		
		0.563	10	0.319		

Energy spectrum of irradiating radiation, is the energy of its constituent particles (gamma and beta) or photons measured as their number as a function of their energy.

Physical and mechanical states refer to the specific activity of the source, aggregate condition, chemical composition of materials from which the radiation sources are prepared, mechanical properties such as strength, viscosity, resistance to thermal stresses and so on and magnetic properties.

In practice, besides the energy spectrum of gamma-radiation of the isotope, the concept of the gamma-constant of the isotope (K_γ) is also used.

It is determined as per the scheme of disintegration and energy of gamma radiation being released. The strength of the exposure dose of the gamma-radiation of a given isotope is termed as the gamma-constant of the isotope and is expressed in units, r/hour, at 1 cm from the source with the strength of unfiltered radiation being 1 m curie. There are differential and complete/total gamma-constants. Differential gamma-constant relates to a definite mono-energetic line of the isotope spectrum. Complete/total gamma-constant is the sum of differential gamma-constants.

^{137}Cs is separated from the gross fission products of nuclear division [137]. ^{137}Cs disintegrates with the release of particles with 0.52 MeV energy, transforms into metastable ^{137}mBa, which in its turn, becomes ^{137}Ba when it releases gamma-photon with 0.661 MeV energy [88]. The stable isotope ^{133}Cs serves as a source for the formation of ^{134}Cs, as per reaction (n, γ), and is one of the most undesirable mixtures of the ^{137}Cs radiation sources, since ^{134}Cs has a very complex spectrum (see Table 5.2) because of the presence of harder rays in the energy spectrum of its gamma-radiation which

Table 5.3. Characteristics of some ^{60}Co and ^{137}Cs sources of gamma-radiation produced by the industry

Overall dimensions, mm		Number of ampules of radiation source	Strength of exposure dose at 1 m from the source, r/sec	Gamma-equivalent, g-eqv. of radium
Diameter	Height			
		^{137}Cs		
11.5±0.2	84–2	2	$(5\pm1.0)\times10^{-2}$	≈20
11.5±0.2	84–2	2	$\left(1.2^{+0.24}_{-0.12}\right)\times10^{-2}$	≈50
11.5±0.2	84–2	2	$(2.1\pm0.42)\times10^{-2}$	≈80
$19^{+0.2}_{-0.3}$	31–2	2	$(1.2\pm0.24)\times10^{-2}$	50
$19^{+0.2}_{-0.3}$	31–2	2	$(2.3\pm0.46)\times10^{-2}$	100
35 ±0.2	48–2	2	$(4.6\pm0.92)\times10^{-2}$	200
35 ±0.2	48–2	2	$(9.3\pm1.86)\times10^{-2}$	400
$38^{+0.2}_{-0.3}$	49–2	2	$(1.4\pm0.28)\times10^{-1}$	600
$38^{+0.2}_{-0.3}$	49–2	2	$(1.8\pm0.36\times10^{-1}$	800
		^{60}Co		
11±0.2	80.5±1	2	4.7×10^{-2}	200
11±0.2	80.5±1	2	1.6×10^{-1}	700
11±0.2	80.5±1	2	3.5×10^{-1}	1,500
11±0.2	80.5±1	2	7.0×10^{-1}	3,000

affects the thickness of protection shield of the units and equipment being designed. The use of ^{137}Cs as a source of gamma-radiation is quite promising because it is based on the separation of isotope from the saline solutions which are obtained as by-products from radiochemical industries and the isotope cost is mainly determined by the technology of its production.

To prepare sources of radiation, we now use cesium chloride (CsCl) and cesium sulphate (Cs_2SO_4), which are obtained by the crystallization method [153, 198], with cesium sedimentation from ammonia or potash by phosphotungstate [196, 360] or with ferrocyanides of nickle or zinc [171, 268]. The specific activity of ^{137}Cs during the preparation of sources can be as high as 25 curie/g [104]. Compression of cesium chloride under the pressure of 1,500 kg/cm^2 increases its density to 3.6 g/cm^3 and makes it possible to increase the specific activity to 90 curie/cm^3. The compression method is used in the USSR.

At present the conveyor production of the gamma-radiation sources with isotopes ^{137}Cs and ^{60}Co is done in the USSR, the USA, England and other developed countries. The characteristics of some of them are shown in Table 5.3. The ^{137}Cs sources used in practice are two hermetically sealed films of stainless steel of the standard X18N10T filled with the ^{137}Cs preparation (Fig. 5.1). At present, the content of the isotope ^{134}Cs does not exceed 6–7% of the activity of the isotope ^{137}Cs. Hermetization of the sources is done by electric welding in a medium of protective gases such as argon. The sources can be used in water-air and neutral media between -60 and $+150$°C.

The isotope ^{137}Cs radiation sources are tested for radioactive surface pollution by the acid extract or smear method.

In the first method, the source is kept in 1–1.5 N solution of nitric acid at 60–90°C for one hour; then the sample withdrawn from the acid is investigated for radioactive substance content. In the second method, a smear is taken with a cotton-wool or muslin tampon moistened with 1–1.5 N solution of nitric acid with the pressure of 0.2–0.5 kg/cm^2 and then the tampon activity is measured.

All sources are marked and are accompanied by special cards showing their details. One of the main characteristics of the source of gamma-radiation is the exposure dose strength expressed in units r/sec and measured 1 m from the working surface of the source. In the serial production of sources, the manufacturing plant uses measuring equipment designed for work in protective chambers and specially graduated for stationary sources of radiation with no more than $\pm 10\%$ error. The strength of the exposure dose created by the radiation sources, whose ratio of height to diameter is more than two, is measured in the direction of the axis perpendicular to the lateral surface of the cylindrical part of the source. As dosimeters (monitors) we use the appliances DIM-60, RP-1M or others in the energy interval of 50–2,000 keV of x-rays and gamma-rays with a detector which has not more than 5% hard radiation filters and not more than 4% reproducibility of readings.

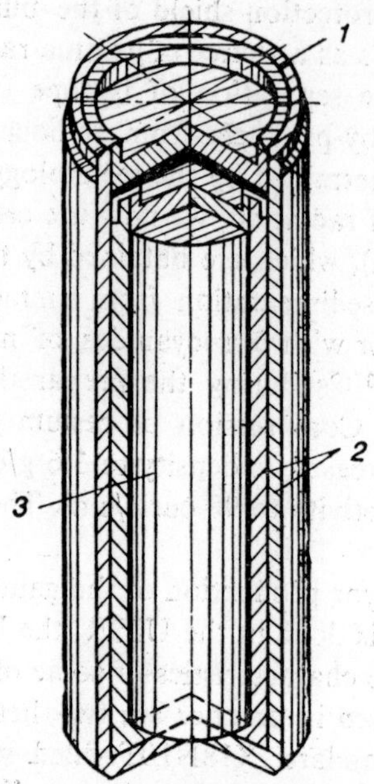

Fig. 5.1. Source of gamma-radiation with isotope ^{137}Cs:
1—argon-arc welding; *2*–ampule;
3—isotope ^{137}Cs.

Although fundamentally the stimulation effect of ^{60}Co and ^{137}Cs gamma-radiation on the seeds of agricultural plants is identical, ^{137}Cs is more extensively used for a number of reasons. For example, ^{137}Cs makes it possible to reduce the protection (shield) thickness to 15–17 cm instead of 26–30 cm for ^{60}Co. The absorbed energy of ^{137}Cs gamma-radiation with a given thickness of the material being irradiated and equal density will be more than for ^{60}Co gamma-radiation under the same conditions, which increases the gainful utilization coefficient of the radiation. The half life of ^{137}Cs makes it possible to recharge the irradiator only after 10–15 years and to obtain reproducible conditions of irradiation, which is extremely important to agriculture. Besides, the full payload of the automobiles or trolleys meant for use on the general network of roads in the USSR (group B) should not exceed 10.5 tons for double axle and 15 tons for triple axle vehicles. This facilitates the use of ^{137}Cs as practically the only source of radiation, whose application allows us not to exceed the above-mentioned parameters as per GOST 9314-59 (the Soviet Standard). As regards the expenditure on acquisition and duration of use of the sources of radiation for gamma-units designed for the presowing irradiation of seeds, ^{137}Cs is comparable with ^{60}Co, particularly if we take into consideration the absence of any expenditure on additional recharging and the smaller construction cost of the unit, including the vehicle for transportation.

In the isotope units for the presowing irradiation of seeds, the cost of ^{137}Cs, as a rule, does not exceed 20–40% of the total cost of the unit. With wider use of gamma-units in practical agriculture and their manufacture at the rate of 100–1,000 units or more per year, the possibility of a further price reduction in ^{137}Cs would make radiation technology even cheaper. Below are the prices of ^{137}Cs in the USA, depending on the activity of the radiation sources being produced.

Activity, curie	Price, dollar/curie
0–1,000	0.5
1,001–5,000	0.45
5,001–2,00,000	0.35
>200,000	0.125

The transportation and storage of the ^{137}Cs and ^{60}Co sources of gamma-radiation is done in special containers. The characteristics of some are given in Table 5.4.

Table 5.4. Characteristics of containers designed to transport gamma-sources

Type of container	Mass, kg	Overall dimensions, mm		Size of box for the source, mm		Place for loading the source
		Diameter	Height	Diameter	Height	
KTB-26-12	2,350	680	860	12.5	85	Canal of the drum
KTB-26-12 "R"	2,350	680	860	14	105	-do-
KIZ-10000	1,230	552	752	40	112	Box of container
KIZ-46	1,630	596	800	28	85	Canal of the drum
KIZ-500	627	434	571	40	105	Removable tumbler

Note: KTB containers are loaded with 1–12 sources.

Transport containers are divided into two groups: the first has containers in which unloading of the sources is done only from the bottom (the containers with the so-called bottom unloading—KTB-26-12, KIZ-46 and others); the second is containers in which the loading and unloading of the sources are done from the top (KIZ-10000 and others) (Figs. 5.2 and 5.3).

BASIC REQUISITES FOR GAMMA-UNITS

Successful application of the radiation technology in agriculture is

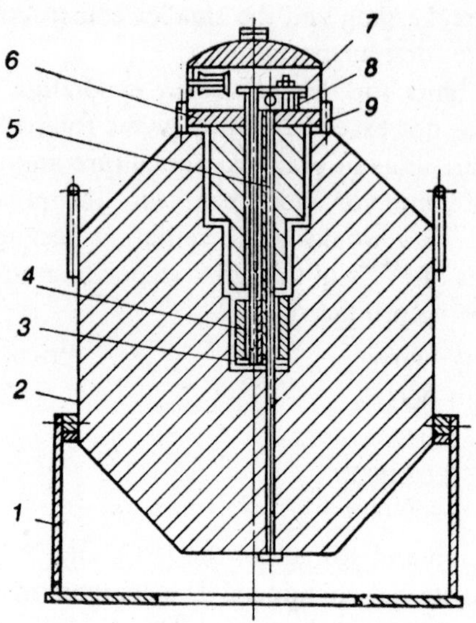

Fig. 5.2. Transport container KTB-26-12:
1—support; *2*—body filled with lead; *3*—layer; *4*—drum with the boxes; *5*—stock or stem; *6*—limb; *7*—fixer; *8*—steering lever; *9*—latch.

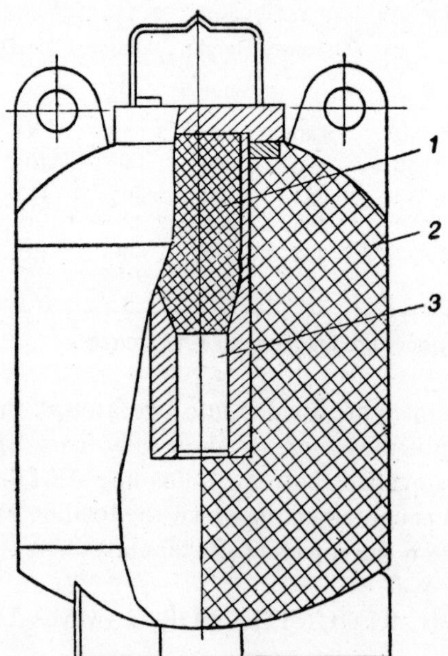

Fig. 5.3. Transport container KIZ-10000:
1—stopper; *2*—body filled with lead; *3*—box for the source.

possible only when it satisfies all the requirements, which depend on the purpose (experimental or commercial gamma-units), conditions of use and properties of the seeds being irradiated.

The main condition for using experimental gamma-units is the provision of wider possibilities to research the presowing irradiation of seeds, i.e., opportunities enabling us to work with seeds which have different properties (size, fluidity, bulk weight, moisture content and so on). The non-uniformity of the irradiation must not exceed $\pm$10–20% and it should permit us to conduct irradiation in a wide range and strength of doses. It is natural that in designing experimental units we must take into consideration the quantity of seeds necessary for sowing under laboratory and field conditions in small plots, competition plots and others. The requirement for this comparatively high degree of uniform irradiation of seeds is easily met by selecting a definite configuration of irradiator, such as a hollow cylinder or a linear or flat irradiator, around which rotate the objects of radiation and maintaining an accurate time exposure for conducting the experiment. The 0.2–1 l volume of the work chamber, as a rule is provided to conduct the experiment, while the 100–800 r/hour strengths of the doses can be applied to investigate the durations of irradiation.

The common prerequisite of all the units is the creation of safe working conditions for researchers and servicing personnel.

One of the main reasons for the limited application of the method of presowing irradiation of seeds until recent years was the absence of commercial gamma-units which could meet the requirements of modern agriculture without changing the already existing technology.

During 1965–1966, in the Special Design Bureau of the Institute of Organic Chemistry of the USSR Academy of Sciences, D.A. Kaushanskii worked out the main prerequisites required from the commercial gamma-units (Table 5.5).

The simultaneous fulfillment of all the requirements in the unit may facilitate the effective application of gamma-units.

Modern and improved design features of gamma-units is a continuous process. Comparatively, minor changes in the design to improve the quality of gamma-units under production must be made with the accumulation of information from the users of the units already in operation and their manufacturers.

The particularly important prerequisite expected from the manufacturer is the homogeneity of the quality of units under production, which will make it possible to increase their reliability and to reduce expenditures on servicing and repairs. Defects may appear even in a unit of very good design as a result of substandard execution during production.

When designing gamma-units, the designer generally tries to meet all the requirements, but invariably he is confronted with contradictions. For

Table 5.5. Essential features required of the commercial gamma-units for the presowing irradiation of seeds of agricultural plants

Branch or enterprise	Requirements
Atomic industry, agricultural machine building, manufacturer-plant	Construction matching with the scales and means of production as well as technological suitability. Use and extensive simplification of units and parts, incorporating design and technological advancements. Conformity of the structural elements of units to the GOST (Soviet standards) and plant standards. Conformity in construction of the irradiator of the gamma-unit to the ideal sources of irradiation. Provision of radiation safety measures during loading and unloading of the radiation sources into the irradiator of the unit. Conformity of all loading operations with sanitation agencies.
Agricultural production, user organizations	*Radio-biological*: Strength of the dose; the degree of non-uniform irradiation; type of the source of radiation; provision of reproducibility of the process of irradiation; provision for applying optimal range of irradiation doses and their control; the storage space for irradiated seeds of different crops before sowing; provision for irradiation of seeds with different levels of moisture content and free-flowing properties. *Agronomical*: Provision of work in different soil-climatic zones; the extent of areas to be sown with irradiated seeds; amount of seeds required for sowing per hectare; physical and mechanical properties of seeds (frictional, composition, size, free-flowing properties, bulk weight, moisture content); type of chemical treatment of seeds; interrelation of parameters of the unit with the existing systems of machines and technological processes. *Operational*: Interrelation of the efficiency of unit with the existing technology and volume of works; strength and durability; non-interruption of work; maximum utilization of the energy of radiation; possibility of control over the dose received by the seeds; simplicity of unit maintenance; ease and convenience of its control; reduced consumption of auxiliary materials; reliability and provision of total radiation safety during operation; reduction of expenditure on servicing and routine repairs; automation of the process of irradiation of seeds and their loading-unloading work; possibility of the prolonged use ot the unit with the sources of irradiation under conditions of agricultural production; adaptability of the unit for movement (self-propelled mobility, transportability).

	Techno-economic indices: Cost of the unit; recovery period of cost or expenditures; expenditure on irradiation; cost of production of agricultural produce as a result of the application of presowing irradiation; per hour cost of seed irradiation or cost of irradiating different types of seeds for sowing per hectare of land; annual economics.
Enterprises producing the sources of radiation	Conformity of the sources of radiation to the construction specifications of the unit and requirements of prolonged operation. Possibility of loading the radiation sources into the irradiation chamber of the unit under manufacturing conditions and their transportation in conformity with radiation safety rules. Reduction of loading time in operation with due consideration for preparatory details; provision of radiation safety.
Foreign trade and international organizations ("Tekhsnabeksport", CMEA and others)	*Clarity of patent*: Increase of competitive ability. Estimate of trade situation and technical requirements of the exporting countries. Production of modified gamma-units suitable to special climatic conditions.

example, to enhance the reliability and durability of a design, it is desirable to increase the purity and accuracy of the conjoining work surfaces, but to do so increases the difficulties of the process and the cost of the product. Therefore, the designer is often compelled to compromise.

The experience gathered on the design and use of a comparatively large number of powerful gamma-units enabled us to frame safety rules No. 482-64 (for units with a mobile irradiator) and No. 774-68 (for units with a stationary irradiator). Basic prerequisites have been worked out in these recognized documents that are required from the gamma-units for radiation safety (see below, for more details).

The breakdown of parts and joints of the unit can lead to failure or even to accidents. Therefore, the durability of parts and joints must be set with consideration for the severest operational conditions. If some of the parts or the procured products cannot withstand the entire amortization period of the unit, then we must envisage and mention their prophylactic replacement or possible repair. However, it must be taken into consideration that good technical and economic indices can be achieved only through the use of gamma-units manufactured under conditions envisaged during the design process.

The commercial gamma-units must provide the possibility of irradiating comparatively large batches of seeds with the given dose and uniformity of irradiation. The designers of the units try to get maximum efficiency with the minimum activity of the irradiator and the smallest unit mass. An important condition for creating commercial gamma-units is the necessity of obtaining a maximum coefficient of use of the irradiator with due consideration for the density of seeds being irradiated and uniformity of the doses of their irradiation. This requirement is governed by a significant (20–40%) percentage of the cost of the ionizing radiation sources (for example, ^{137}Cs) in the total cost of commercial gamma-units

Since the commercial gamma-units, as a rule, work on a large area and in different agroclimatic zones where it may be necessary to sow seeds immediately after irradiation, the units must have self-propelled mobility.

The use of commercial gamma-units under the conditions of collective and state farms, where material losses are possible because of idle time, stipulates the need for reliable units. Reliability of gamma-units naturally implies the overall properties of joints and parts on which depends the performance during the irradiation of seeds to the given indices during their entire use. Some of these properties are described below.

The stability of the process of executing irradiation of products, i.e., the property of all systems of the unit to accomplish the working process without breakdown.

Design resistance of the sources of radiation and materials from which they are prepared to such forces of destruction as corrosion, vibration and several other factors.

Resistance of the design elements of the irradiator, conveyors and blocking mechanisms to destruction from corrosion, radiation, environmental conditions, vibration, fatigue, wear-and-tear and the material strength.

Stability of physical and chemical properties of materials, i.e., their capacity to maintain geometric shape and physico-mechanical properties under given conditions.

Faultlessness of design and manufacture.

The unconditional requirement of technical documentation for gamma-units is the inclusion of technical and economic bases of the development with each redesigned unit.

MOBILE COMMERCIAL GAMMA-UNITS

The first gamma-unit developed especially for conducting research on different aspects of the method of presowing irradiation of seeds of agricultural plants under laboratory and field conditions was the transportable unit GUPOS [45]. It was developed at the Institute of Biophysics of the USSR Academy of Sciences. Between 1966–1969, it was manufactured under the name Stebel. Some technical data of this unit are given below:

Source of radiation	^{137}Cs
Efficiency of irradiation in the dose of 1,000 rad, kg/hour	20–25
Cumulative maximum activity of irradiation, curie	2,100
Average strength of dose in the working volume of the chamber, r/min	700
Degree of nonuniformity of the dose field, %	± 20
Volume of the working chamber, liter	1
Strength of dose on the surface of the unit container, mr/hour	2.8
Total mass of the unit, kg	6,400

The Stebel unit was mounted on the 2PN-4 autotrolley which was transported by a ZIL-130 automobile (Fig. 5.4).

A number of stationary gamma-units are also used in our country for research: GUBE-800 [46], GUBE-4000 [44, 116], Gidroponika [63], LMB-gamma-1M [120], VIESKh [148], UGU-200 [48] and others. Among foreign units, the stationary gamma-unit Rotatron is of some interest and is described in a paper [356] and uses ^{137}Cs as the source of radiation.

In 1966–1967 a mobile commercial gamma-unit Kolos was designed and constructed (Fig. 5.5). It made it possible to irradiate seeds in a continuous flux [124, 125].

Figure 5.6 shows the general scheme of the Kolos unit. The unit moves itself to one of the sections of the farm, where the servicing personnel

Fig. 5.4. Portable gamma-ray unit Stebel.

Fig. 5.5. Mobile commercial gamma-unit Kolos
(experimental model).

bring it to working condition and load the seeds in the receiving bunker *7*, they go through the scoop conveyer *2*, into the dose receiving bunker *1* of the irradiation block *3*. Passing through the working chamber containing an assortment of linear sources in the form of a hollow cylinder with a linear irradiator in the center, the seeds enter belt conveyor *6* and from there to the bunker *9* and finally to the bag *10*. The unit is controlled from the control desk *5*. The irradiated seeds from the bag are poured directly into the seed drill immediately after irradiation. The unit can be used in practically any region of the country and can move on all motorable roads.

The basic design of the irradiation block is shown in Fig. 5.7. The unit irradiator comprises an axis part *8* in the cavity of the shaft *5* which is situated along vertical axis of the unit and parts representing an assortment of linear sources *3* along the cylindrical surface of the squirrel wheel. Such an

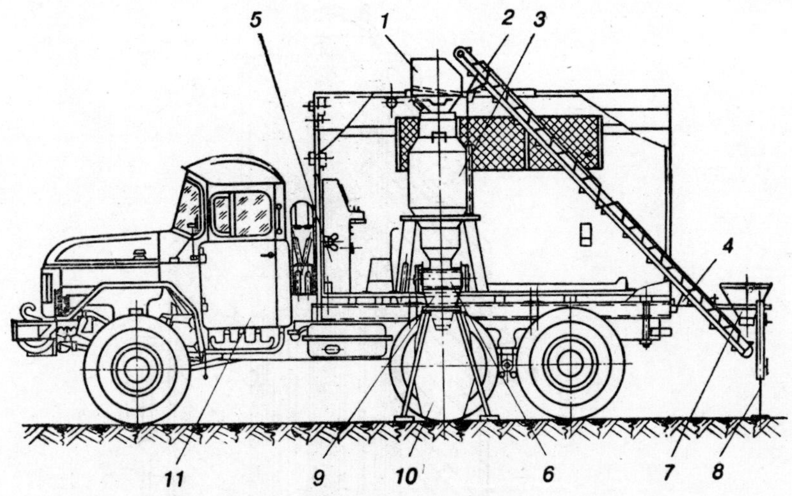

Fig. 5.6. Scheme of the mobile commercial gamma-unit Kolos:
1—Dose receiving bunker; *2*—scoop conveyor; *3*—irradiation block; *4*—attachment of the scoop conveyor; *5*—control desk; *6*—belt conveyor; *7*—receiving bunker; *8*—support of the receiving bunker; *9*—bunker; *10*—bag; *11*—ZIL-131 automobile.

arrangement of sources makes it possible to increase the coefficient of radiation utilization with due consideration to such important factors as a constant dose strength and uniform irradiation. For example, in the presowing irradiation of seeds, the strength of the dose must be within a range of 800 rad/min, because much higher dose strength leads to yield decrease. The sources have casings *3* that protect the seeds against overdose, in the direct vicinity of the irradiator. ^{137}Cs is selected as the radiation source.

Input *5* and output *10* canals contain protective elements, one part of which *4* is embedded in the shaft *1* and the other *6* in the protective body *7* forming the circular clearance of the technological canals *5*, and *10*; through these canals the poured material is mixed by the action of its own weight. This design of the technological canals makes it possible to construct units with less mass and provide complete radiation safety.

In the design of output canal *10* the dosing mechanism *11* is made in the form of a saucer-like valve which is fixed longitudinally by the control mechanism *12*; it enables us to reduce or increase the circular space of the output canal. In order to avoid the formation of "bridges", the possibility of rotating the central shaft is envisaged with the electric drive *15* and mechanical transmission *14*. The work chamber with irradiator is built in the protective body *7*.

The creation of the Kolos unit made it possible to launch extensive commercial trials of the presowing irradiation of seeds in the MSSR during

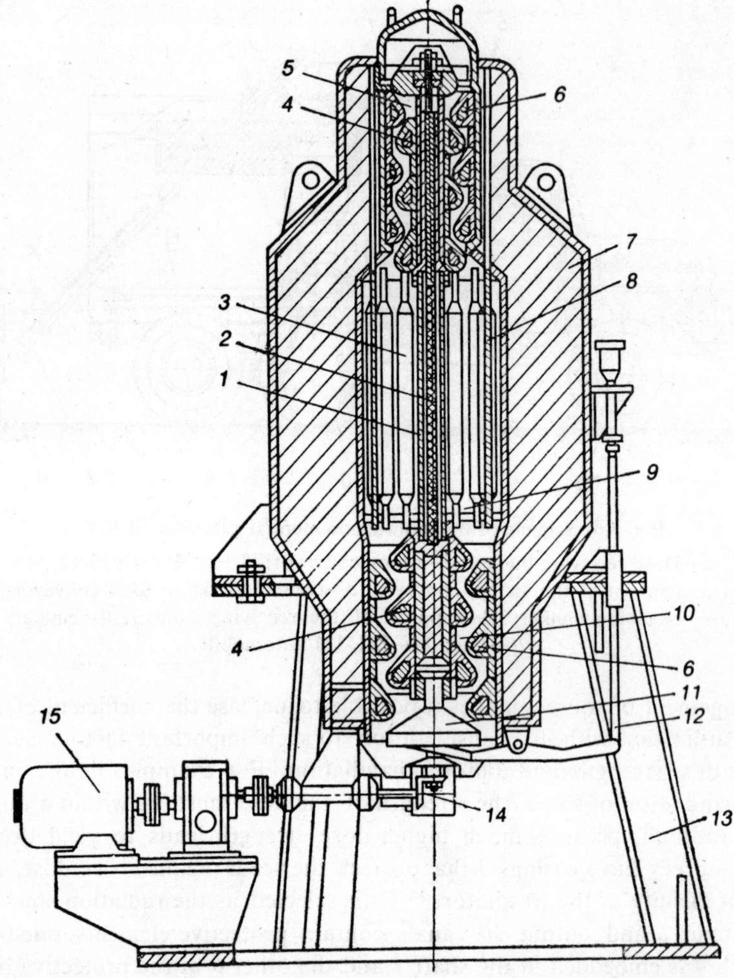

Fig. 5.7. Irradiation block (in cross section) of the gamma-unit Kolos:

1—shaft; *2, 8*—^{137}Cs radiation sources; *3*—cassette of the irradiator cylinder with protective casing; *4, 6*—protective elements; *5, 10*—input and output technological canals; *7*—protective body; *9*—working chamber; *11*—dosing mechanism; *12*—control mechanism of the dosing mechanism; *13*—mount or frame; *14*—mechanical transmission of rotation of the shaft; *15*—electric drive.

1968–1970 and its practical application in farms during 1971–1973 [118, 119]. During the commercial trials of the Kolos unit, its design and operational qualities were tested against the requirements of modern agricultural production and its radiation safety was investigated. During 1971–1973 the manufacturer plant brought out nine serial Kolos units (Fig. 5.8) and ten units in 1974.

Fig. 5.8. Mobile commercial gamma-unit Kolos (serial specimen).

Simultaneously, as a result of three years of trials on the Kolos units, its design was modernized with the objectives of reducing the time required in rolling and unrolling the transport mechanism of the unit, improving the general working conditions of service personnel and reducing the contact time with different seed-dressing chemicals (Hexachloran, Granosan and others) [122]. If 30–40 minutes were required to bring the experimental model unit to working conditions with simultaneous inputs of physical forces earlier, then perfecting the transport mechanisms reduced the volume of auxiliary work and reduced the preparation time for the unit to three to five minutes. Simultaneously, the service personnel's contact time with seed-dressing chemicals was also reduced considerably following the hermetization of the feeding conveyor. The feeding conveyor moves with about 10–15 kg seeds. The scoop conveyor is enclosed in the hermetic casing.

In the process of designing the commercial gamma-unit Kolos, a huge volume of engineering and physical calculations were done to select the best design and geometry of the irradiator, determine the dosing fields as well as the protective (shield) thickness against ^{137}Cs gamma-radiation, with due consideration to the extent of the irradiator and multiple scattering. Technical characteristics of the unit are shown in Table 5.6.

Radiation parameters verified during trials of the Kolos units led to the following: the efficiency of 1 ton/hour is achieved with irradiation in the dose of about 700–800 rad; the degree of non-uniform irradiation does not exceed the stipulated values.

In order to verify the estimated parameters of irradiators, the dosing field calculations were done by several methods [129]: engineering physical methods based on the concept of shaft sources, in the form of superposition of the point sources (on the Minsk-22 computer); engineering methods based on the concept of the squirrel wheel in the form of superpositions of the

Table 5.6. Basic technical data of gamma-units

Name of indices	Kolos	Universal	Stimulator
Efficiency of irradiation in the dose of 1,000 rad, kg/hour	750–1,000	500	4–5
Degree of non-uniformity of irradiation, %	±20	±20	±20
Average strength of dose in the working chamber, r/min	800–1,000	400	800
Source of radiation	^{137}Cs	^{137}Cs	^{137}Cs
Activity of irradiator, curie	3,500	4,300	720
Coefficient of gainful use of radiation, %	17	8–10	4
Requirements toward:			
average bulk weight, g/cm³	0.5–0.8	up to 1.5	up to 1.5
average moisture content, %	10–12	Any	Any
size of the seeds, mm	5–15	1–100	1–15
free-flowing property	Free flowing	Any	Any
The nature of technological process of irradiation of seeds	Continuous	Cyclic	Cyclic
Range of change in the irradiation doses, rad.	300–5,000	100–100,000	100–100,000
Factor determining the change of radiation dose	Change of the flow of seeds	Change of the duration of irradiation	Change of the duration of irradiation
Strength of dose on the surface, mr/hour:			
radiation block	⩽2.8	⩽2.8	⩽2.8
hood	⩽0.28	⩽0.28	⩽0.28
Type of drive	Electric	Electric	Electric

Supply from electric generator mounted in the hood of the unit or from autonomous source of A.C. current in voltage, V	200 or 380	220 or 380	—
Required strength, KWt	1	2	—
Type of automobile or trolley	Automobile ZIL-1	Automobile KrAZ-2558	Automobile GAZ-69 or trolley GAZ-704
Mass of the unit without automobile (trolley), kg	3,500	6,000	500
Mass of the unit with means of transport (complete mass), kg	10,300	16,000	1,850/850
Servicing personnel, number of persons	3	3	1

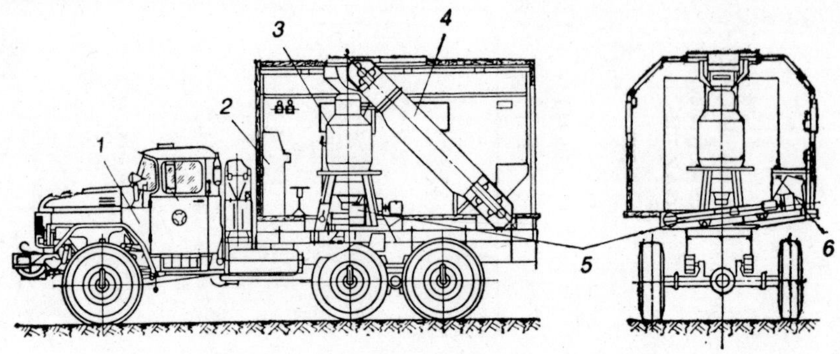

Fig. 5.9. Schematic diagram of the gamma-unit Kolos after modernization: *1*—ZIL-131 automobile with hood; *2*—control desk; *3*—radiation block; *4*—feeding conveyor; *5*—selecting conveyor; *6*—autonomous built-in electric generator.

linear sources with uniformly distributed activity (on the Minsk-22 computer and manually) and engineering method with combination of the Monte Carlo method (on the Minsk-22 computer). Results of the calculations are presented in Table 5.7. The first three methods give results practically concurrent with each other (Fig. 5.10).

Strength of the exposure dose (relative unit) is linked with the strength of the dose (r/min) by the equation: R (r/min) $= 6.4 \times 10^2\ r$ (relative unit).

Physical constants of radiation, such as coefficients of weakening and true absorption, were given for some effective energy, which was estimated as the average energy of spectrum at maximum output from the irradiator. Energy spectrum was calcuated by the Monte Carlo method, with due consideration to the real design of the irradiator. Below are the energy spectra at maximum output from the shaft irradiator of squirrel wheel type of the Kolos unit (for ^{137}Cs; Eo $= 0.661$ MeV; $\bar{E} = 0.55$ MeV):

Interval of energy, MeV	0.661–0.611	0.611–0.561	0.561–0.511	0.511–0.461	0.461–0.411	0.411–0.361
Number of particles in the given interval	3.631	162	175	161	188	192
Interval of energy, MeV	0.361–0.311	0.311–0.261	0.261–0.211	0.211–0.161	0.161–0.111	0.111–0.061
Number of particles in the given interval	194	231	263	215	91	13

Despite the significant difference of the average spectrum energy from the initial energy, the coefficients of true absorption change insignificantly: from 0.0327 to 0.0332 cm^{-1} for ^{137}Cs, which practically does not affect the dose distribution within the working volume of the unit.

Table 5.7. Some estimated experimental results of verification of radiation parameters of the gamma-unit Kolos

Size, cm				Efficiency, ton/hour	Time of irradiation, min	Gamma-equivalent, g-eqv. of radium		Average volume strength of dose, rad/min	Uniformity of dose field, %	Coefficient of use of radiation, %	Integral dose, rad	Remarks
Chamber		irraditor										
Diameter	Height	Diameter	Height			central shaft	squirrel wheel					
27.5	43.0	22.5	42.0	0.67	1.0	500	1,100	950	±20	11.0	950	Considering the volume of irradiator
27.5	43.0	22.5	42.0	1.07	1.0	500	1,100	920	±20	17.0	920	
27.5	43.0	22.5	42.0	1.0	—	450–550	1,100	—	±18	16	850	Average experimental data of 10 units

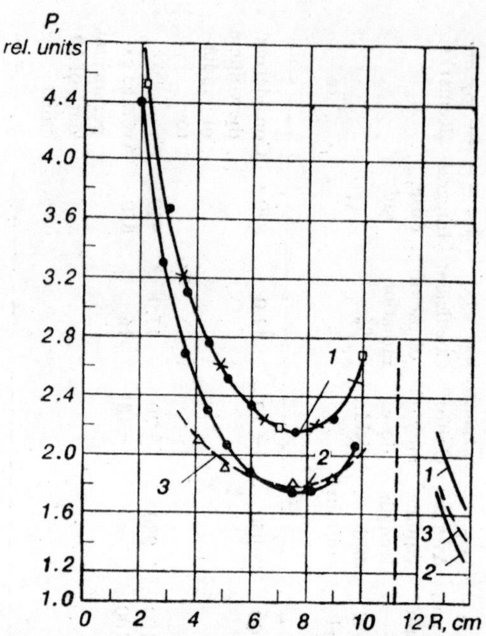

Fig. 5.10. Distribution of the strength of dose (relative unit) along the radius in the middle plain for the gamma-unit Kolos:

1—in the air: ●—in point proximity, ×—in linear proximity (on the Minsk-22 computer), □—in linear proximity (manual calculation); 2—in the water equivalent medium with density of 0.7 g/cm³ (point proximity); 3—experimental data (medium—corn).

The degree of the non-uniformity of the dose field, as obtained through calculations, is significant but, if we consider that the sharpest fall of dose function is observed in the immediate vicinity of the sources, then we may say that in a major part of the volume (in 88% of the volume) the uniformity of the dose field comprises the given value of ±20% (Fig. 5.11). It must be mentioned that in the actual unit, some equalization of the dose field occurs with the help of design elements (see Fig. 5.8) which "intersect" the portions of the working volume, deteriorating the irradiation uniformity.

Figure 5.12 shows the estimated isodose field of the working chamber of the Kolos unit. Practically, as a result of investigations the degree of non-uniformity of irradiation of seeds did not exceed ±18% on three Kolos units manufactured by the industry and ±12% on six other units. The efficiency of the units was approximately one ton/hour with irradiation in doses of 700–800 rad.

The coefficient of radiation utilization, equal to the ratio of the energy

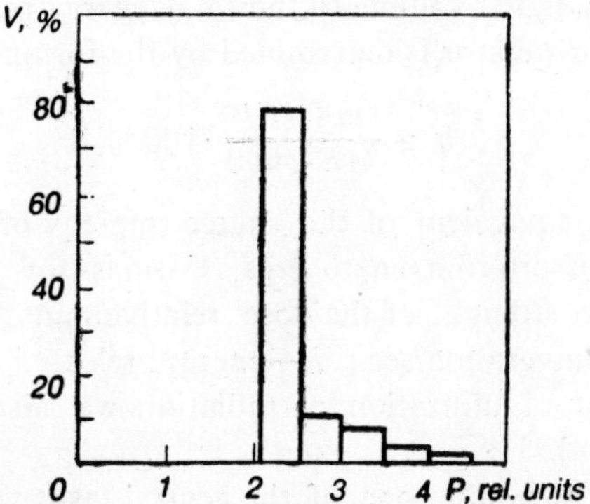

Fig. 5.11. Histogram of strength distribution of dose R (relative units) through the volume V in the gamma-unit Kolos.

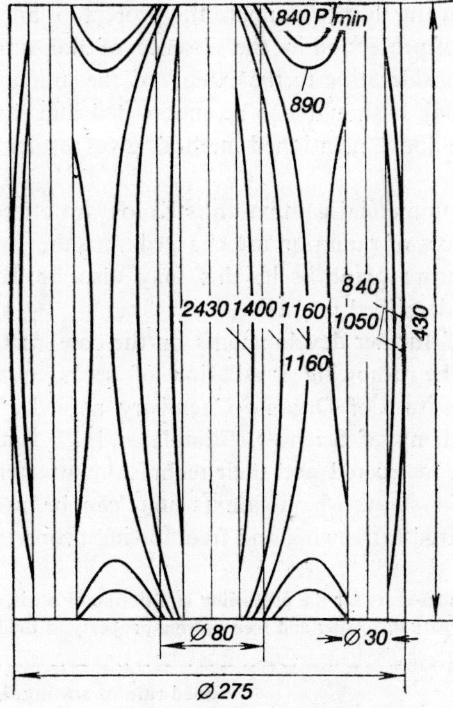

Fig. 5.12. Isodose field of working chamber of the gamma-unit Kolos.

absorbed in the irradiated volume to the total energy released by the radiation source of the irradiator is determined by the formula:

$$\eta = \frac{mK_\gamma fVP}{AE \cdot 3600} \cdot 100\%,$$

where, m—gamma-equivalent of the source (mg-eqv of radium); f—coefficient of conversion from röntgen to ergs; V—mass of irradiated seeds, g; P—average volume strength of the dose, relative units; A—total activity of the irradiator, disintegration/sec.; E—energy released per disintegration, erg. The coefficient of utilization of radiation was also determined by the formula given in [44].

While building the unit, one of the central tasks was the estimate and optimization of protection against gamma-radiation. It was necessary to obtain a minimum mass of protection and simultaneously solve the problems associated with continuous supply and selection of the flow of seeds into the radiation chamber. Spectrometric and dosimetric characteristics of technological canals in the unit to estimate the protection are given below.

The estimate of protection in the absence of non-uniformity in it was done with due consideration to the extent of the source and multiple scattering [44, 105, 106]. It should also be mentioned that the estimate of protection, as per the above-mentioned method, gives results with a reserve of about 10%.

The mobile commercial gamma-units Kolos are offered with the source of radiation. Sources of radiation are charged into the protection chamber at the manufacturing plant itself; this may also be done by the "dry" method, which is described in [121].

As a result of further developments on the construction of commercial gamma-units for the presowing irradiation of seeds in the special design bureau of the Institute of Organic Chemistry named after N.D. Zelinskii of the USSR Academy of Sciences, Stimulator [117] and Universal [119, 123] units were constructed and their technical characteristics are given in Table 5.6. Table 5.8 shows which gamma-units can be used, depending on the seed-rate required for sowing and free-flowing property of the seeds.

Table 5.8. Gamma-units for the presowing irradiation of seeds, depending upon seed-rate of sowing and free-flowing property of the seeds

Type of seeds	Seed rate of sowing, kg/ha		
	0.1–3	5–100	150–300
Not free-flowing and poorly free-flowing	Stimulator	Universal	—
Free-flowing	Stimulator	Kolos	Kolos, Kolos-5

During 1970–1971, the design of the mobile commercial gamma-unit Universal was completed. It is meant for the presowing irradiation of seeds of different agricultural plants that do not flow freely, for example, cotton and cuttings of vegetatively propagated plants (Fig. 5.13).

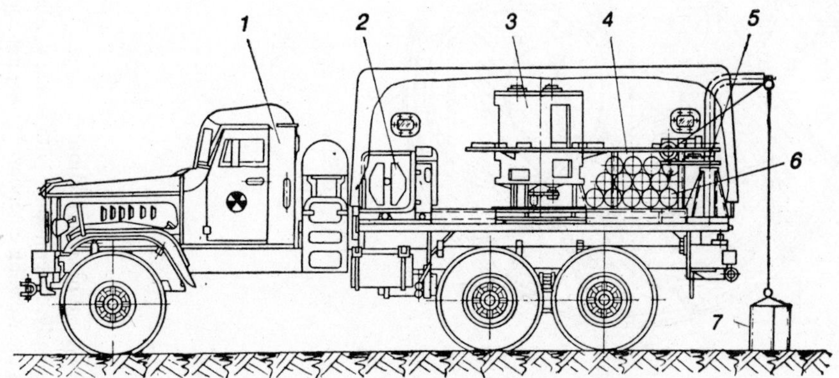

Fig. 5.13. Mobile commercial gamma-unit Universal:
1—KrAZ-255B automobile; *2*—autonomous electric generator; *3*—radiation block; *4*—distributor table; *5*—lift mechanism; *6*—bag; *7*—container.

In Fig. 5.14, *a* and *b* show the schematic diagram of the design of the radiation block in the form of a welded body; block *2* is filled with lead, in which three vertical canals are situated at an 120° angle (see Fig. 5.14, *b*). In each canal, the rotors with work chamber *9* are mounted on ball-bearing supports. Along the central axis of the radiation block there are canals to route the irradiator cassette with radiation sources *1*. In the upper part of the radiation block is a protective lid *10*. Inside the working chamber, a turning mechanism *8* is mounted on the vertical shaft which has the container *3* with seeds prepared for irradiation. At the time of irradiation the disk with the container is rotated around the irradiator with an electric drive *7*, at a speed of 5–7 rpm. Turning the rotors by 180° from the position "Object for irradiation" to position "Object out of irradiation" and vice versa is done with electric drive *6*, which is in the lower part of the radiation block. All three rotors with work chambers are turned simultaneously. The container-bag to be filled with seeds or other objects for irradiation is a cylindrical pot with a capacity of 19 liter; a rotating distributor table *4* is at the level of lower edge of the work chamber of the body of the radiation block. The radiation block along with the rotor drive is mounted on the frame *5*, which in turn is fixed to a special frame on the automobile. Ideal sources with radioisotope ^{137}Cs are used as the source of radiation.

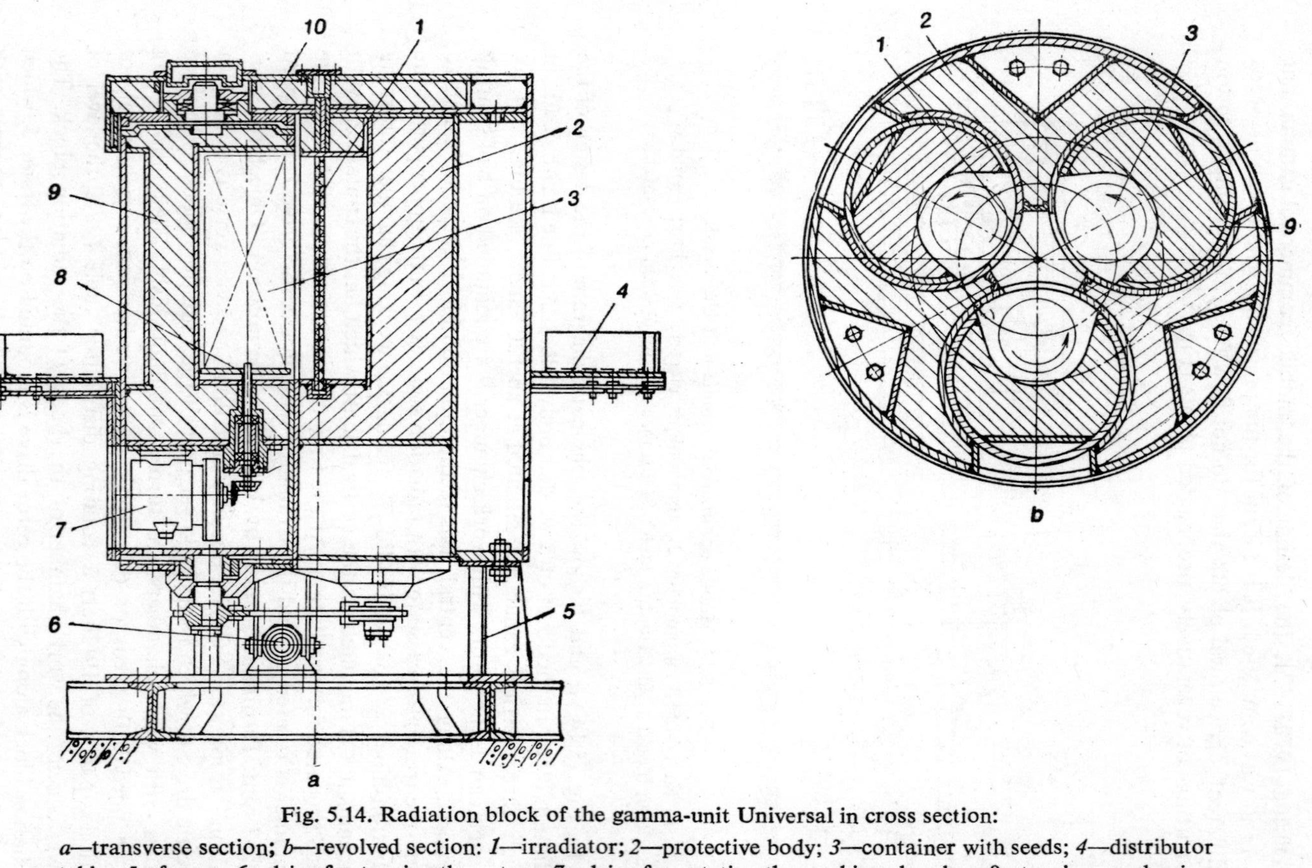

Fig. 5.14. Radiation block of the gamma-unit Universal in cross section:
a—transverse section; *b*—revolved section: *1*—irradiator; *2*—protective body; *3*—container with seeds; *4*—distributor table; *5*—frame; *6*—drive for turning the rotors; *7*—drive for rotating the working chamber; *8*—turning mechanism of the working chamber; *9*—rotor with working chamber; *10*—protective lid.

The working principle of the gamma-unit Universal stipulates that seeds or other biological objects (for example, cuttings of vegetatively propagated plants) are put into the cylindrical containers, six of which are placed in the conveyor container and fed onto the working table by the lift mechanism. The operator places the containers manually along the roller conveyors of the table in the special sections of the distributor turn table. By pressing the button "Load," the operator brings rotors with working chambers into the original position "Object out of irradiation," loads the seed filled containers into work chambers and by pressing the button "Irradiation", brings the rotors to the work position "Object for irradiation;" the rotation of containers with seeds in the chambers begins simultaneously with switching on the rotors. Because the dose field and strength of the dose are determined experimentally before starting the operation of the unit, the given irradiation dose is achieved by varying the time of exposure. A time relay is provided to make the work easy and after a definite period of time, brings the rotors with irradiated seeds to the position "Object out of irradiation."

The irradiated seeds from the containers are manually removed from the chamber and fed onto the work table and new containers with unirradiated seeds are loaded into the chamber for irradiation.

Using the unit to irradiate potato tubers may be considered an experimental-cum-commercial unit to conduct extensive trials under field and large-scale commercial cultivation conditions. In this case its efficiency is 1–1.5 tons/hour for irradiation in the dose of 300 rad. Voropaev and associates [75] report the possibility of designing a mobile gamma-unit to irradiate seed potato with an efficiency up to 10 tons/hour.

In 1969 a mobile gamma-unit named Stimulyator was designed and constructed (Fig. 5.15). The experimental model of this unit was charged with ^{137}Cs radiation source in March, 1970 and put into operation. The unit is designed to irradiate seeds of agricultural plants with small sowing seed-rate (0.05–3 kg/ha), as well as for work on varietal trial farms, testing stations, in research institutes and on experimental farms. For industrial crops like tobacco, it can even be used as a commercial unit.

The unit (Fig. 5.16) comprises a detachable container (body *5* and lid *6*). In the center of the body of the container, there is an immobile cylindrical irradiator in the form of a cassette *2* of six tubular cells with sources of radiation *1*. Sources of radiation of ^{137}Cs isotope are ideal with an 11 mm dia and 84 mm height. The irradiator is charged in the protective chamber or under water. Protective stoppers *3* are put into holes of the canals after charging and are hermetically sealed by electric welding in the medium of protective gases. Along the axis of the container in a directing tube, a stock or piston *4* is placed with a work chamber. The stock with work chamber is placed around the irradiator by the manual lift mechanism *9*. The work chamber is lifted and lowered manually. The unit is mounted on a GAZ-69

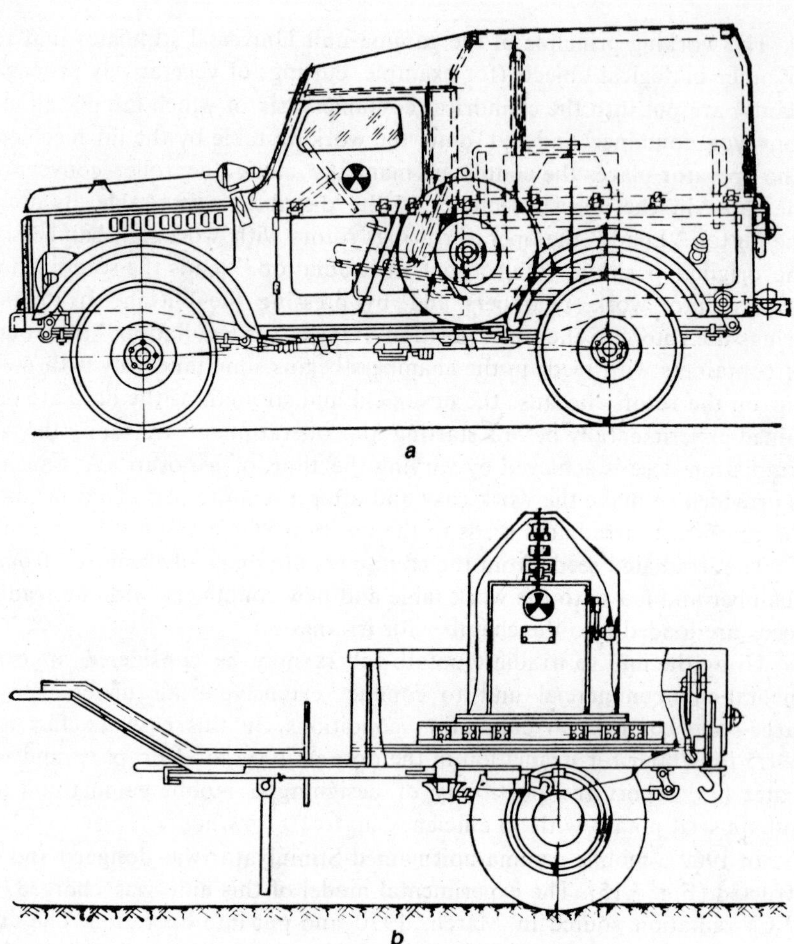

Fig. 5.15. Mobile gamma-unit Stimulator:
a—**on the GAZ-69 automobile;** ***b***—**on the trolley.**

automobile or on GAZ-704 trolley (see Fig. 5.15) and is protected against atmospheric precipitation by a special hood.

The unit is operated in the following manner: seeds are poured into special 0.2 liter volume tumblers which are put in the work chamber and then placed in radiation by a mechanism. One person is needed to operate the unit.

The irradiator of the experimental model of the unit (Fig. 5.17) is charged with six ^{137}Cs sources of radiation with a total activity of about 720 curie.

The surveys conducted by the Institute of Nuclear Energetics of the Academy of Sciences of BSSR, showed that the strength of the dose on the

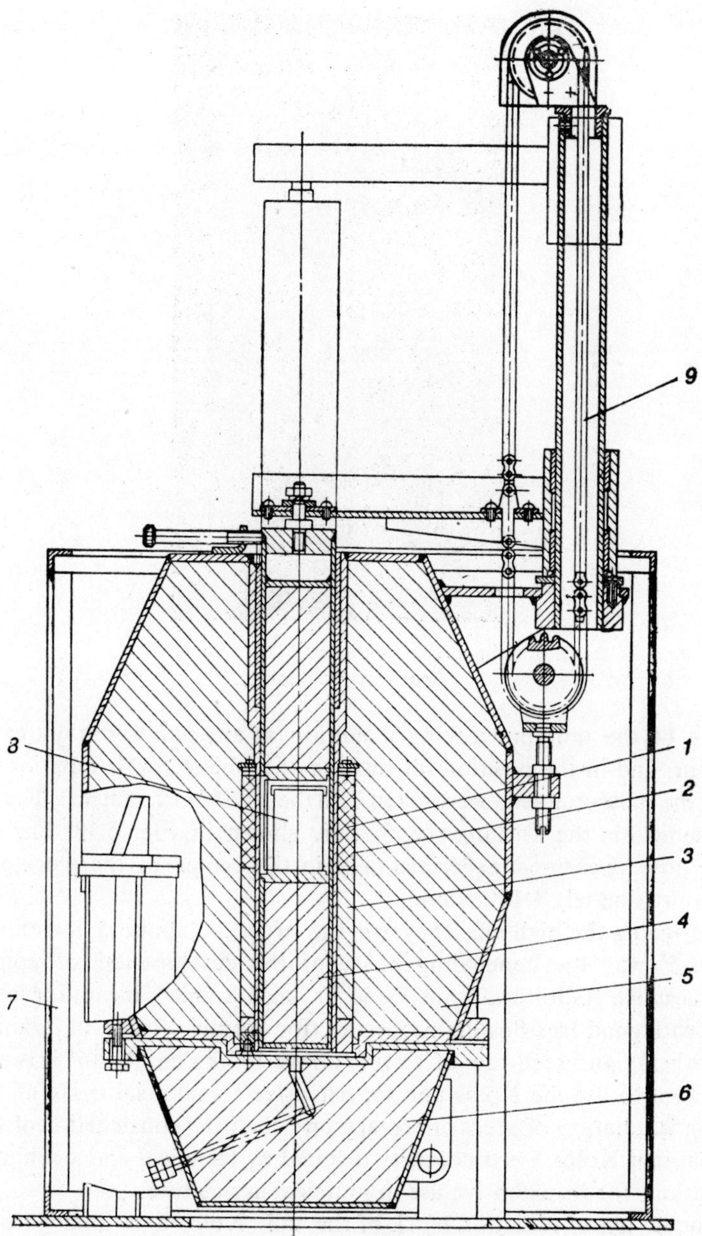

Fig. 5.16. Unit 'Stimulator' (radiation block in cross section): *1*—sources of radiation ^{137}Cs; *2*—irradiator cassette; *3*—protective stopper; *4*—stock or piston with work chamber; *5*—protective body; *6*—protective lid; *7*—decorative framework; *8*—tumbler of the work chamber; *9*—lift mechanism (lowering).

Fig. 5.17. Gamma-unit 'Stimulator (experimental model).

surface of the unit in working condition at any point does not exceed 0.5 mr/hour, and in the position for loading the object for irradiation—2 mr/hour (the measurements were taken with the RUP-1 instrument). The changes of dose field in the working chamber are shown in Fig. 5.18. The strength of the dose, measured experimentally in the center of the work chamber, was approximately 1,100 r/min.

Designing the high efficiency (to the order of about 5 tons/hour) unit 'Kolos-5' was the concluding phase of the development of commercial gamma-units. As follows from Table 5.8, such a unit is meant for irradiating seeds with good free-flowing properties (for example, seeds of grain cereals and others) and seed-rate of sowing above 100 kg/ha. And if it is adequate at present to use the Kolos unit for extensive commercial trials of the pre-sowing irradiation of seeds of cereal crops, then the construction of the commercial unit Kolos-5 is a necessity dictated by technical and economic considerations for its extensive use in practical agriculture.

The design of radiation block of the unit Kolos-5 is analogous to that of the radiation block of the Kolos unit (see Fig. 5.7). Here ^{137}Cs is also selected as the source of radiation.

The following are the estimated parameters of the Kolos-5 unit: efficiency—5 tons/hour with irradiation in the dose of 1,000 rad; degree of non-uniformity of irradiation—$\pm 20\%$; mean volumetric strength of the dose in

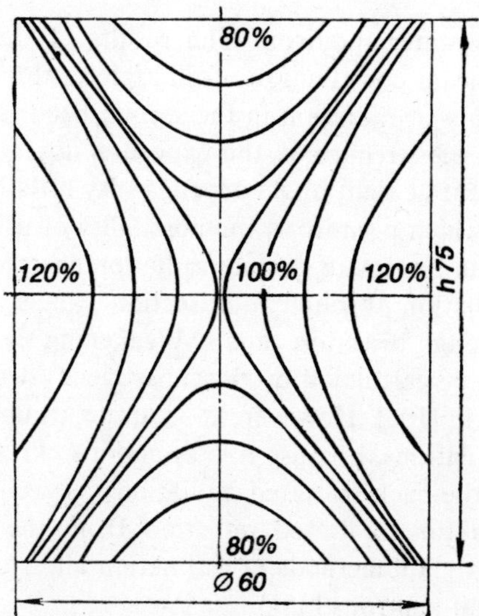

Fig. 5.18. Dose field of the gamma-unit 'Stimulator'.

the medium with density of 0.7 g/cm^3—about 1,000 rad/min; coefficient of utilization of radiation—15%; total activity of irradiator—about 20 curie.

The Kolos-5 unit, similar to the Kolos unit, makes it possible to irradiate seeds in a continuous flow.

With the launching of industrial production of a whole complex of mobile commercial gamma-units by our industry, it will be possible in a short while to put into practice the presowing irradiation of the seeds of agricultural plants as a new agronomic measure.

CHARACTERISTICS OF GAMMA-RADIATION PASSING THROUGH MULTISTEP CANALS OF GAMMA-UNITS FOR SAFE CONTINUOUS SEED IRRADIATION

When designing commercial gamma-units with continuous irradiation of seeds in a flow, one of the main problems is the estimate or experimental determination of optimal protection against radiation with the simultaneous provision of an uninterrupted supply of seeds into the work chamber and their continuous selection. The protection complex comprises technological canals which make it possible to provide a continuous supply of seeds into the work chamber and their selection. Simultaneously there is a provision to reduce the strength of the dose from beyond protection to permissible norms. There are fairly numerous detailed research reports [195, 271, 272, 316, 340, 359] on weakening gamma-radiation in rectangular, cylindrical and canals bent at right angles passing through a protection made of cement, concrete, aluminum water and iron. The results of these research reports point out the significant contribution of multiple scattered gamma-radiations toward total intensity of radiation in the second bend (section) of the canal. The distribution of the strength of the exposure dose of the point isotope sources in canals bent at right angle depends very little on the energy of the gamma-radiation, which permits us the possibility of using the data obtained experimentally and estimate gamma-radiation energies [359].

Passage of radiation through multisection canals has been studied in less detail than through bisection canals. Weakening the radiation in multisection canals can be calculated by the subsequent estimate of each bend as per methods given in [195]. However, it is quite troublesome and difficult to make such calculations, because it is essential to know the spectral-angular distribution of the back-scattered radiation, as well as to consider the contribution of multiple-scattered radiation from the walls of the canals. The existing approximate methods of estimation allow us to solve this problem only for thermal neutrons [351].

To estimate the passage of radiation through multistep circular canals, a method has been proposed which is described in [316]. However, this method has not been confirmed experimentally.

In order to fill this void, D.A. Kaushanskii, E.D. Chistov and O.F. Partolin [128, 249–251] investigated the passage of gamma-radiation of point isotope sources ^{137}Cs and ^{60}Co with an activity of 5 m curie through a multistep technological canal (Fig. 5.19), which forms the input mechanism model of the gamma-unit Kolos. The source was placed at different distances from the inlet of the canal, which made it possible to model the volumetric irradiator with the following measurements: diameter 190 mm, height 120 mm. The input mechanism model was made of steel and covered with a detachable steel or lead protective shield, depending on the experiment.

As a detector of gamma-radiation, the SBM-10 meter with filters to reduce the energy and severity [161, 179] was used with a PP-12 conversion mechanism. Spectral characteristic of the scattered gamma-radiation was studied at different points in the canal with a scintillation spectrometer, in which the detector was a sodium iodide crystal (NaI) (T1) measuring 40 × 50 mm and a FEU-56 photomultiplier.

During the measurements, the SBM-10 meter was placed at different points along the axis of the canal, whose positions are shown in Fig. 5.19. Measurements were also taken in the multistep canal filled with the material to be irradiated (wheat, corn, sunflower and beet seeds) and in the canal filled with air. Maximum relative error during the measurement of the gamma-radiation intensity did not exceed 3%, but when measuring its spectral distribution it was 10% [250, 251].

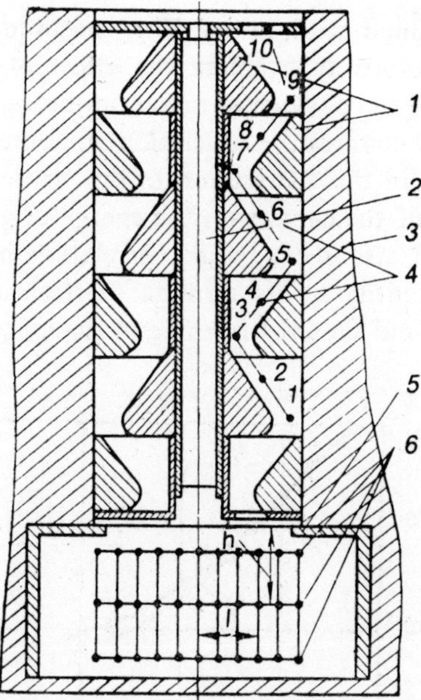

Fig. 5.19. Multistep technological canal in the protection of the gamma-unit Kolos (full-scale model):

1—protection sections; *2*—shaft; *3*—biological protection (lead); *4*—coordinates of location of the detector; *5*—body of the chamber; *6*—coordinates of the source of radiation.

The results* of measuring the intensity of scattered ^{137}Cs gamma-radiation along the axis of the canal filled with wheat seeds are shown in Fig. 5.20. The diagram shows the correlation of the ratio of ^{137}Cs gamma-radiation intensity at the point under study along the axis of the canal to the radiation intensity at a point immediately after the first bend in the filled canal. The distribution of ^{137}Cs gamma-radiation intensity along the axis of the empty or hollow canal is also given for comparison. As we see, in the beginning, the gamma-radiation intensity changes very little, but then we observe a drastic fall and finally it equals. The shape of the weakening intensity curves of gamma-radiation along the axis of the canal depends very little on the energy of radiation; this is particularly noticed at terminal portions of the curves. At each point along the axis of the canal, the relative intensity of scattered gamma-radiation is greater in the canal filled with air than in the canal filled with wheat seeds. At points on the outlet of the canal, the relative intensity of scattered gamma-radiation reduces by 30%

*Analogous results were obtained when the canal was filled with seeds of corn, sunflower and beet.

when the canal is filled with grains. The measurements so recorded make it possible to quantitatively evaluate the effect of filling the canal with moving grains during the passage of radiation through the canal.

Data in Fig. 5.20 confirm the conclusion presented in [272, 359] about the weak correlation of the distribution of the scattered radiation intensity after the first bend of the canal with the energy of gamma-radiation. From this diagram, we can also see a weak correlation between the weakening gamma-radiation intensity at its terminal portion and the position of the radiation source around the inlet opening of the canal.

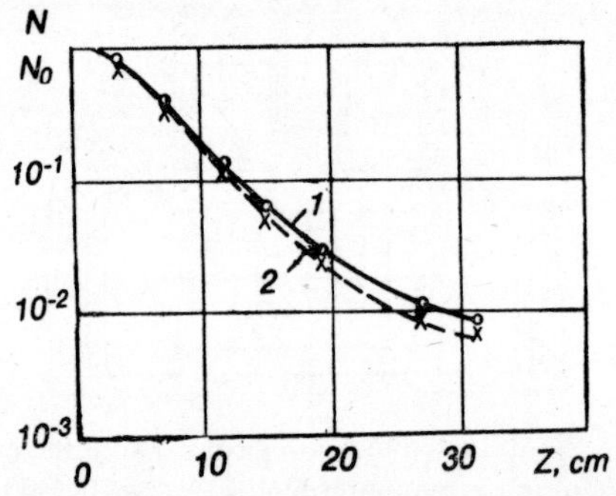

Fig. 5.20. Relative distribution of scattered ^{137}Cs gamma-radiation intensity along the axis of the empty or hollow (*1*) and filled (*2*) canals in different positions of the radiation source.

O: *h*—2 cm, *r*—4 cm; ×: *h*—12 cm, *r*—4 cm.

The distribution of scattered gamma-radiation intensity along the axis of the canal can be approximately described by the following empirical equation [128, 250]:

$$I_z = I_0 c z^{-k},$$

where, I_z and I_0 represent the radiation intensity at point z on the axis of the canal and at the point after the first bend of the canal respectively; c and k are empirical constants depending on the geometry of the canal and the material of the protection shield.

To study multistep circular steel canals with a protective shield made of lead, the values of coefficients, depending on the geometry of canal and material of the shield, comprise 3 and 2.6 respectively [128]. The value of the coefficient k for the above-mentioned protection shield material practically does not depend on the energy of gamma-radiation, relative size of the canal and the angle of its bend for intervals of changes of the studied parameters. The value of the coefficient c depends on the material of the protection shield, energy of radiation and angle of the canal bend.

It must be mentioned that conversions are used when determining the gamma-radiation intensity from sources of different types. They stipulate that the strength of the dose from the source of a certain type is determined as per estimate or experimental data for intensity from other type of source [106]. So, according to this particular work [185], if a cylindrical source (radius 10 cm, height 30 cm) is replaced with a point source situated in the geometrical center of the cylinder, the maximum error in the determination of intensity at points on the axis of the canal after the first turn is 25%, but for the points after the second turn the error will not exceed 15%. For subsequent points along the axis of the canal, the error with such replacement will be still less. In this way, we may admit that with known values of error the distribution of scattered gamma-radiation intensity along the axis of a multistep canal from the cylindrical and the point source in the geometrical center of the cylinder will be identical.

The results of investigations on the spectral composition of gamma-radiation passed through a multi-sectional canal are shown in Figs. 5.21 and 5.22. Spectral distribution of scattered gamma-radiation in the low-energy range of the spectrum is maximum around 250 KeV, corresponding to the multiple scattering of gamma-radiation. In the region of energies of gamma-radiation of the ^{60}Co source, we observe that the maximum spectral distribution is caused by unscattered gamma-radiation. When the canal is filled with wheat seeds, apart from reducing the cumulative intensity of gamma-radiation, there occurs a qualitative change of the spectrum of the scattered radiation, particularly with an increase in the relative contribution of multiple scattered gamma-radiation accompanied by small displacement of the maxima of spectral distribution in the region of 100–120 KeV, i.e., there is a definite softening of the spectral radiation which is coming from the canal. The shape of the spectral distribution curve practically does not change with the placement of the source of radiation at different points in the irradiator. This situation is stipulated by the fact that the spectral distribution of multiple scattered gamma-radiation at the outlet of the canal attains an equilibrium, i.e., the ratio of intensities of individual spectral groups becomes constant irrespective of the initial energy of the radiation and the location of the source around the inlet opening of the canal.

When applying the ^{137}Cs source, the contribution of unscattered gamma-radiation toward the total radiation intensity becomes noticeable only if the number of turns (bends) in the canal is reduced, i.e., through the removal of one or several sections (see Fig. 5.21, curves *3*, *4*). In such a case, the ratio changes between the contributions of unscattered and the multiple scattered gamma-radiation toward the total radiation intensity at the outlet from the canal. Thus, with reduced turns (bends), the proportion of unscattered gamma-radiation increases and under certain conditions (see Fig. 5.21, curve *4*) the relative contributions of the multiple scattered and unscattered gamma-radiations toward the total gamma-radiation intensity at the outlet of the canal become measurable.

The spectra of scattered gamma-radiations shown in Figs. 5.21 and 5.22 were obtained for when radiation sources were placed at the following coordinates: $h = 12$ cm, $z = 4$ cm. With the placement of the radiation source at other points in the irradiator, the shape of the spectral distribution curves of radiation practically does not change. This condition is stipulated by the fact that spectral distribution of multiple scattered gamma-radiation at the outlet from the canal attains equilibrium, i.e., the ratio of the intensities of individual spectral groups becomes constant, irrespective of the initial radiation energy and location of the source vis-a-vis the inlet opening of the canal.

The experimental data obtained about the weakening gamma-radiation in a multistep, circular canal in the protection shield of the gamma-unit

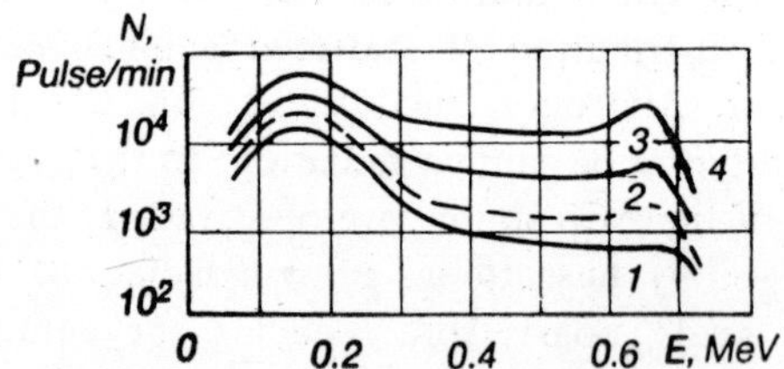

Fig. 5.21. Spectral distribution of ^{137}Cs gamma-radiation at the outlet from the canal:
1—maximum collection of sections; *2*—one section removed; *3*—two sections removed; *4*—three sections removed.

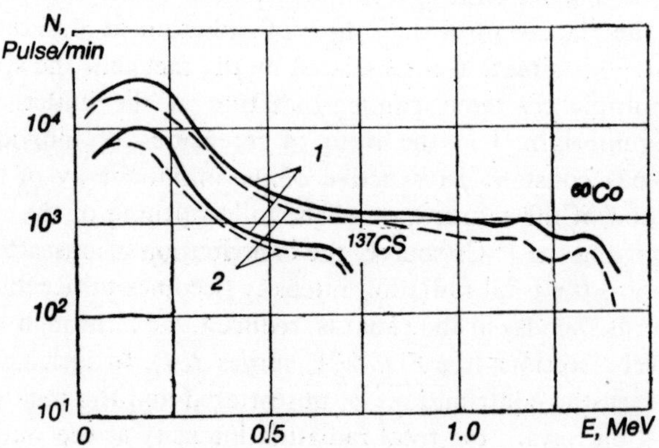

Fig. 5.22. Spectral distribution of ^{137}Cs and ^{60}Co gamma-radiations at the outlet from the empty (hollow) (*1*) and wheat filled (*2*) canals.

enabled us not only to give quantitative and qualitative assessments of the intensity of scattered radiation passed through the canal (with due consideration to filling the canal with and without the scattering medium), but also to develop realistic definite gamma-unit designs for the continuous irradiation of seeds.

DOSIMETRY OR MONITORING PRODUCTION PROCESSES

Monographs [5, 114, 192, 197, 254, 256], textbook [112] and quite a large number of original works [92, 223, 231] are devoted to the problems of dosimetry. We dealt with only those dosimetry methods which may be applicable in conducting experimental works on the large-scale commercial utilization of the presowing irradiation of seeds.

The task of dosimetry in the presowing irradiation of the seeds of agricultural crops is to determine the radiation dose received by a particular type of seeds being irradiated and its control. Dosimetry is based on the registration of physical effects (ionizing, thermal, chemical, optical and others) which determine the radiation energy absorbed by the object being irradiated.

As a rule, the radiobiological effect is predetermined by the absorbed radiation energy. Here we must take into consideration the fact that the radiation dose is equal to the product of the strength of the dose and time and is measured by the radiation energy absorbed by a unit mass of the medium. Such a dose is called the absorbed dose and is measured in rads (1 rad $= 10^{-2}$ joule/kg of the absorbing mass).

The problem of measuring the absorbed dose can be solved with different methods [5, 92, 112, 114, 192, 223, 254, 256]. For application in the presowing irradiation of seeds the dosimetric systems must work in the dose range of 10^2–10^4 rads, not depend on a change in the gamma-radiation energy within the range of 0.2–3 MeV and strength of dose in the interval of 1–25 r/sec and also temperatures between -10 and $+35$°C, be resistant to the effects of light, mixtures and oxygen, make comparisons relatively easy, have a measurement error within $\pm 10\%$ and, above all, be as cheap as possible.

Depending on the technological process of seed irradiation—cyclic or continuous—the problems of measuring the absorbed dose are solved by different methods. For gamma-units based on the use of the internal cavity of the cylindrical irradiator, in which the container with seeds is placed and in which, at the time of irradiation, the object is either immobile (Stebel'-3A, Stimulator) or rotates around the stationary irradiator (Universal), the problems of dosimetry are solved comparatively easily. For this purpose, the dose field is investigated by one of the available dosimetric methods. For example the method of chemical dosimetry (ferrous sulphate method or any of its modifications to measure a low dose) or the method of thermoluminescent dosimetry. The ferrous sulphate method is often used for graduating other dosimetric systems but, in the USA, it is even used as a standard method to determine the absorbed energy of both gamma- and x-ray radiation [263].

The measurement and control of the absorbed dose in the continuous process of irradiation in the flow of seeds puts forth a number of additional requirements: the size of detectors must be commensurate to the size of seeds being irradiated; the density of the detector must be close to the average density of the seeds being irradiated; detectors must have adequate mechanical strength besides being capable of storing the information. Therefore, for dosimetry in the continuous flow, it is recommended to use thermoluminescent detectors at the base of aluminophosphate glass, activated manganese or lithium fluoride [126, 127]. This method is based [254] on the fact that the gamma-radiation energy absorbed by the detector is stored in it and then is registered in the form of visible luminescence released upon heating. Since the photo current of thermoluminescence is proportional to the absorbed dose of ionizing radiation, we can determine the absorbed dose of the ionizing radiation by measuring it.

For gamma-units with cyclic irradiation, it is recommended to use the ferrous sulfate method for the certification of work chambers. However, lately the thermoluminescent method has been used quite frequently. In order to get uniform measurements of the absorbed and exposure doses, the authors of the work [50, 76] had worked out methodical instructions for determining the conduct of state certification of dose fields in the work chambers of gamma-units with immobile irradiators [166].

We shall examine in detail the above dosimetric methods.

The ferrous sulfate method* is based on the principle that under the action of ionizing radiation in water solutions of ferrous sulfate containing sulfuric acid and sodium chloride, the bivalent ferrous ion is determined spectrophotometrically at a wave-length of 304 nm. Regarding atomic composition and density (1.024 g/cm^3), the dosimetric ferrous sulfate solution is close to the composition and density of many species of seeds. At doses of 3×10^2–4×10^4 rads, dose strength to 10^7 rads/sec, temperature 0–50°C and energy of gamma-radiation up to 2 MeV, the radio-chemical output of ions $\pm e^+$ per 100 eV of absorbed energy comprises 15.3 ± 0.3 ions. The molar coefficient of extinction of the trivalent ferric ion in the ferrous sulfate solution at 20°C and a wave length of 304 nm comprises $2{,}097 \pm 3$ l/(mol·cm). The dose can be determined from the following relation:

*Methodical instructions for the ferrous sulfate dosimeter have now been worked out in the All-Union Research Institute of Physical and Technical Measurements of the State Committee of Standards of the USSR Council of Ministers.

$$D = \frac{2.914 \times 10^4 (S - S_0)}{1 - 0.0065 (20 - t) l} \text{ rad,}$$

where S and S_0 are the optical density of the irradiated and unirradiated solutions at 304 nm respectively; l—the thickness of the layer (measuring trough, cuvette or cell) cm; and t, the solution temperature at the time of measuring the optical density.

The optical density of the solution is measured in 1 cm wide quartz cuvettes (for particularly accurate determinations the cuvette should be 2 cm wide at the wave length of 304 nm and using 0.4 M solution of sulfuric acid as the standard solution on the spectrophotometer. The same temperature is desirable, otherwise a 0.65% per 1°C temperature correction is applied in the above formula. Ampules with the solutions are brought to thermal equilibrium before determining their optical density.

Generally salts and acids are used for the solution, viz., a double salt of ferrous oxide, ammonia and sulfuric acid ("analytically pure" as per Soviet standard GOST 4208-66, 4204-66). It is recommended to filter out lime from the acid by passing it through a No. 3 glass filter. The distilled water used must also not contain any soluble inorganic or organic impurity. It is therefore recommended to use double or even triple distilled water. Good results are obtained from a ferrous sulfate solution of the following composition: in 5 l of distilled water we dissolve 2 g of $FeSO_4 \times 7\ H_2O$ or 2.8 g of $Fe\ (NH_4)_2 \cdot (SO_4)_2 \cdot 6\ H_2O$, 0.3 g of NaCl and 110 ml of 95–98% H_2SO_4. If the solution is in contact with air for some hours, it is not necessary to additionally saturate it with oxgyen. The solution can be stored up to one month in a dark glass vessel.

Ampules in which the solution is irradiated for the series of measurements must be of similar geometry and of the same material. The ampules are not hermetically sealed, but are left open. However, if it is necessary to measure in closed ampules, it is necessary to saturate the solutions with oxygen well in advance. The ampules can be prepared from chemically resistant glasses (pyrex, neutral glasses), quartz, teflon and stainless steel. The ampules of stainless steel must be treated with 96% nitric acid well in advance [232]. Most often glass or teflon ampules are used. Before use they are cleaned with a mixture of potassium dichromate and sulfuric acid in the ratio of 1 : 20 (chromic acid) and finally rinsed 3–4 times with tap and distilled water and with the working solution. The ampules must have a volume of 3–5 ml and a diameter not less than 12 mm. Doses of 7×10^3–3×10^4 rad enable us to get the most accurate measurements. While passing oxygen through the solutions, it is recommended to connect the gas cylinder with tubes made of polymeric materials and cleaned of dust. We must avoid the use of rubber tubes and stoppers for reproducible results, because they lead to organic contamination.

The spectrophotometers used must have passed through corresponding tests, because the error of measuring the coefficient of extinction of trivalent ferric ions on different spectrophotometers can be significant in the ferrous sulfate method.

The relative error of determination of the absorbed dose with the ferrous sulfate dosimeter is calculated by the formula:

$$\delta D = \sqrt{\delta G^2 + \delta\epsilon^2 + \delta n^2},$$

where δG—relative error of the determination of radio-chemical output of Fe^{3+} with the help of colorimeter (error of the certification of standard detector), 6%; $\delta\epsilon$—relative error of the determination of coefficient of molar extinction, 2–3%; δn—relative error of readings on the measuring instrument, 1%.

The ferrous sulfate method is used to measure and certify dose fields of gamma-radiation in the work chambers of the gamma-units designed for experimental and commercial use in the following cases: before making the unit ready for operation; additional loading or some another reason for a change in the activity or geometry of the irradiator (for example, in case of a complete or partial replacement of the sources of radiation); and periodically with changing experimental conditions. Results of dosimetrical investigations are incorporated in the technical data sheet of the unit or in the special certification document. The dose received by the object is determined during further irradiation.

The small size of the detectors, strength, resistance to weather conditions and the possibility of comparatively quick processing of information about the absorbed dose with $\pm 10\%$ error, are some of the factors that make it possible to consider thermoluminescent dosimetry with aluminophosphate glasses activated by manganese or lithium fluoride as detectors as a promising practical method [126, 127]. Some prerequisites of dosimeters for the presowing irradiation of seeds in continuous flow are given in Table 5.9; it is but natural that the list of such prerequisites may be even longer.

In our country, as well as abroad, thermoluminescent dosimeters have been developed according to specification to measure doses on gamma-units for the presowing irradiation of seeds and their control. We shall examine their use in the commercial gamma-units Kolos [126].

Russian-made thermoluminescent dosimeters DTM-2 and IKS as well as the VA-M-30 made in the German Democratic Republic can be used in these units.

Some technical specifications of these instruments are shown in Table 5.10.

Aluminophosphate glasses with manganese used as a detector are prepared in the form of polished rectangular sheets measuring $10 \times 3 \times 3$ mm or circular disks with a diameter of 8×1 mm. The thermoluminescent glow

Table 5.9. Dosimeter prerequisites for the presowing irradiation of seeds

Prerequisites	Parameters	Remarks
1	2	3
No correlation with the energy of gamma-radiation	from 70 KeV to 1.5 Mev	
Linearity of the measured parameter in the interval of doses	10–14^4 rads	
No correlation with the strength of dose for gamma-radiation	up to 25 rad/sec	
No correlation of sensitivity with:		
Light	Storage in the diffused light must not change the dose parameters	Before irradiation, at the time of irradiation and in the process of storage
Temperature	—10–+50°C	
Technological specificity in the preparation and absence of the effect of admixtures		Difference in response of one batch of detectors should be distinguished from the other by not more than ±5%
Reproducibility of results within a range of	±3%	Determined by instruments in the laboratory and state trials
Error of the measurement of absorbed dose, not more than	±10%	The use of a graduated scale is permissible. It is recommended to make graduation under conditions close to parameters of the commercial gamma-units

Contd.)

Table 5.9 (*Contd.*)

1	2	3
Mechanical strength	Not less than 2 kg/cm^2	With the use of LiF powder, the concept of strength relates to the capsule
Overall size	Not more than 10 cm × 3 cm × 3 cm	
Tissue equivalence or water equivalence	—	
Stability of indices after irradiation	To three months	Here we mean the duration of storage with the absence of any post-radiation effects
Technological specificity of application during use	—	Reduction of time/labor consumption
Low cost and possibility of serial production	—	Here we mean the issue or distribution price
Competitive abil ity and fabrication to patented specifications	—	—
Production of modified detectors suitable for use under special conditions	—	The conditions are agreed upon during the development
Conformity with the scales and means of production including technological specificity of preparation	—	—

Table 5.10. Basic technical data of thermoluminescent dosimeters

Name of indices	DTM-2	VA-M-30	IKS
Range of measurements of dose, r	$10–10^4$	$1–10^4$	$10^2–10^4$
Size of detectors, mm, not more than	10×3×3	dia. 5×15	dia. 8×1
Time required to measure the dose accumulated by one detector, minutes; not more than	3	1	1/6
Time during which the detector stores information about the absorbed dose, in months (with storage in darkness and not above 35°C), not more than	6	—	6
Total error of determination of dose, %	±10	±15	±7
Instrument warm up time after switching it on, minutes; not more than	30	30	5
Power supply to the instrument for A/c of frequency 50 cycles/min. volts	127/220	220	220
Power requirement of units, W; not more than	350	60	50
Total weight of the instruments, kg; not more than	14.5	30	4

Note: Detectors for DTM-2 and IKS—of aluminophosphate glass with manganese, for VA-M-30—of the photocomposition VA-S-200.

is measured in a special heating mechanism of the thermoluminescent dosimeter. The luminescence intensity is registered by a photomultiplier and the photo-current in the chain of a photomultiplier is measured by a microammeter or with a direct current self-recording instrument.

Thermoluminescent detectors make it possible to fix the absorbed doses within a range of $10–10^4$ rads, but the energy accumulated in the glasses or lithium fluoride does not depend on the strength of the dose of the ^{137}Cs and ^{60}Co gamma-radiation to 25 rads/sec [126]. Any abrupt manifestation is practically absent in the energy range from 70 KeV to 1.25 MeV. The characteristics of thermoluminescent detectors were verified for 0.66 MeV energy of ^{137}Cs gamma-radiation with due consideration for the specific conditions under which the monitoring was done.

Detectors can be used a number of times and they are capable of storing information for several months, which is quite important under field production conditions. The instruments used must be certified by the standards institutions.

Dosimetric measurements for the mobile gamma-unit Kolos are done in two stages.

The laboratory stage in the experimental verification of radiation parameters of the unit, determination of dose-chart statistics for different crops, preliminary testing of the control mechanism and so on. As a result, the actual efficiency of the unit in a definite dose (for example, 1,000 rads) and degree of non-uniformity of irradiation of the seeds are fixed.

The production stage is the registration of the exact dose received by the seeds and the compilation of graduated tables in the first and second year of operation with different crops of the given geographical zone, with due consideration to composition of the batch of seeds and the treatment method (for example, seed-dressing). The results are endorsed and incorporated in the technical data sheet of the unit. Later, the monitoring is done periodically, because the dose is determined by the consumption of the irradiated seeds (by the farms for commercial cultivation) and is established in accordance with the table for the given crop.

Additional monitoring is done periodically to check the working of the control mechanism. For this purpose, the detectors are wrapped in black paper and three or four detectors are put into the receiving bunker, where they pass through the work chamber along with the flow of grain. They are then taken out manually and measurements are done as described above. Each experiment is repeated three to four times. The average dose is worked out on the basis of three or four experiments. Then the error of determination of the degree of uniformity of seed irradiation is calculated, a new seed rate is determined and the measurements are repeated as per the same method.

The thermoluminescent dosimetry method can be used not only for the presowing irradiation of the flow of seeds of cereal, pulse and vegetable crops, but also when the interval of doses is 10–10^4 rads and the energy spectrum lies within 70 KeV to 1.5 MeV. It can also be appropriately used in the radiation treatment of potato before planting, for which the detectors should be placed in the midst of the tubers, which must be marked. These tubers should also pass through the same cycle as other irradiated tubers. Then in any direction of tubers we can determine the dose from the dose-chart statistics, which is particularly important for the irradiation of potato in heaps rather than in bags. We thus get the real phantom—the detector records of the dose inside the tuber.

The practical application of one or two methods of dosimetry enables us to not only measure and control doses, but also to obtain uniform measurements.

TECHNOECONOMIC EFFECTIVENESS OF THE PRESOWING GAMMA-IRRADIATION OF SEEDS AND WIDER APPLICATIONS OF RADIATION TECHNOLOGY IN AGRICULTURAL PRODUCTION

Experience from operating commercial gamma-units Kolos has shown that, along with technical and technological advantages, they have high economic indices and they recover their cost in the very first year of use in terms of increased yield and quality of the produce. Technical and economic indices of the effectiveness and wider use of the new radiation technology and new agronomic measure of the presowing irradiation of seeds are of prime importance in practical agriculture, in that they play a decisive role in envisaging their wider application. In all events, the method of estimating the economic effectiveness of the new radiation technology of presowing irradiation of seeds as a new agronomic measure is based on several works [9, 19, 64, 165, 167, 229, 230].

To determine the economic effectiveness of the presowing irradiation of seeds with the new radiation technology used for commercial cultivation (for crop husbandry), the following basic indices are taken into consideration:

the gross increase in the harvest of crop per hectare of planted area of the farm, region, province and so on in terms of natural and cost input expressions;

increase in the output of marketable produce in quintals and their monetary value in rubles;

output of additional produce per unit rubles spent in connection with the use of radiation;

increased net profit (cost of the produce minus additional expenditure) per hectare sown with irradiated seeds or per unit rubles of expenditure;

increased profitability (ratio of the net profit to expenditure, %);

change of the cost of production of agricultural produce;

effectiveness of additional capital investments in gamma-units, characterized by the duration of their return (ratio of expenditure to annual sum of net profit), by the coefficient of economic effectiveness (the value opposite to the value of recovery) and other indices.

To determine the effectiveness of the new agronomic measure and the new technology, we take into consideration the gross and marketable volume of agricultural production in the cost assessment. Gross marketable produce is assessed on the state and collective farms on the basis of purchase prices and the marketable produce—on the basis of the actual sale proceeds in the farm.

Assessing the effectiveness of the presowing irradiation of seeds only on the basis of increased marketable produce cannot be accepted as a decisive index for all crops, because the marketable produce reflects only one part of the volume of agricultural production. For example, if the marketability of produce is 100% for cotton and tobacco, then for cereal crops during the last five years it was less than 50% and only a few percent for fodder crop.

The presence of commercially cultivated areas sown with irradiated and unirradiated seeds of agricultural crops (moreover, the seeds can be irradiated in different doses, in different climatic zones and sown at different times) under identical agroclimatic conditions of cultivation make it possible to determine more accurately the economic effects of the application of new technology and the method of presowing irradiation of seeds, on the basis of the data on the increased yield and net or gross profit (without consideration for the overhead expenditure) through irradiation with the optimal stimulation dose.

Overhead expenditure is determined as a percentage of the expenditure on labor on the farm (or on salaries) linked with the application of the radiation technology.

Net profit on the farm from the use of radiation technology is determined by the formula:

$$N = C - E,$$

where N—net profit, in rubles; C—cost of the produce obtained as a result of the use of radiation technology, in rubles and E—expenses linked with the use of radiation technology, in rubles. These expenses can be shown by the components:

$$E = A_1 + A_2 + A_3 + A_4 + A_5 + A_6 + A_7,$$

where A_1—depreciation deductions on the renovation of the gamma-unit during its operation for 15 years; A_2—expenses on current repairs of the unit; A_3—salaries of the service personnel of the unit; A_4—transportation expenses (without labor payment), which include expenditures on the technical maintenance of the automobile, wear-and-tear of tires, lubrication, fuel, materials and so on; A_5—expenditure on fuel to operate the unit; A_6—expenditure on harvesting, processing and transportation of the additional produce; A_7—expenditure on sale of the additional produce obtained as a result of the presowing irradiation of seeds.

We must take into consideration the fact that the half life of ^{137}Cs, which is used in practically all commercial gamma-units, is almost twice the expected life of the unit and, therefore, a decrease in the efficiency of the gamma-unit beyond the life period of the unit is comparatively small ($\approx$2.4% in a year) and the transfer of the required cost to the ready produce may be considered equal.

The norms of depreciation deductions from different types of major heads have been differentiated, depending on the conditions of work. On the irradiation unit, the annual norm of depreciation can be taken as 10%, in analogy with the norms applied to the equipment of atomic power stations [19] or more accurately—taking 15 years as the period of operation of gamma-units [165].

Evaluation of the produce obtained as a result of the application of radiation technology can be done on the basis of prices at the actual sale, as well as on the basis of the cost, depending on the objective of analysis or investigation. In the comparative determination of the effectiveness of the means of irradiation used in different farms (regions, provinces, etc.), the prices must be comparable within limits of the region or province.

While studying individual marketable crops on the state farm or collective farm (for example, cotton or tobacco) the sale prices are of great interest because they more accurately reflect the actual quality of the produce. The yield of fodder crops is assessed in purchase prices or through the cost of fodder units on the basis of forage grain.

Increased net profit obtained from the application of radiation technology (in comparison with the profit obtained from the area sown with unirradiated seeds) is determined as per formula:

$$N_0 - N = G_0 = (T_3 + E) - (G - T_3) = (G_0 - G) - E,$$

where N_0 and N—net profits from the area sown with irradiated and unirradiated seeds respectively, in rubles; G_0 and G—gross produce from the area sown with irradiated and unirradiated seeds respectively, in rubles; and T_3—total expenses on the cultivation of the crop except expenses on irradiation, in rubles.

One of the indices of economic effectiveness of the radiation technology can be the recovery of expenditure, which is determined as per formula:

$$T_{re} = C_{ea}/N_p,$$

where T_{re} – period of recovery of capital investment, in years; C_{ea} – capital expenditure on acquisition of the unit, in rubles; N_p – sum of net profit in rubles from the additional produce from all crops whose seeds were obtained and sown in one year before and after irradiation in the unit.

Profitability (P) of the use of radiation technology for presowing irradiation of seeds and that of the agronomic measure is determined as per formula:

$$P = \frac{C - E}{E} \cdot 100\%.$$

Changes of the cost per unit production of the produce obtained as a result of the application of presowing irradiation of seeds can be determined with the following formula:

$$D = E_3/G, \; D_0 = (E_3 + E_0)/(G + G_0),$$

where D and D_0—cost per unit produce of unirradiated and irradiated seeds respectively, in rubles; E_3—all expenses incurred per hectare (without expenses on irradiation), in rubles/ha; E_0—additional expenses on irradiation; G_0—increase of yield obtained as a result of the application of presowing irradiation of seeds, q/ha.

Labor efficiency in getting the additional produce as a result of the irradiation of seeds before sowing is determined using the formula:

$$El_0 = G/G_0, \; El = (G + G_0)/(G_0 + Al_e),$$

where, El_0 and El—efficiency of labor on areas sown with irradiated and unirradiated seeds respectively, in quintal/man-hour; Al_e—additional labor expenses on the application of presowing irradiation of seeds, in man-hour per hectare.

If the farm fields have control areas, then the economic effectiveness of the presowing irradiation of seeds is determined in the same way as for field or large-scale commercial trials under the conditions of research institutes and farms. In this case, the indices obtained from the area sown with irradiated seeds are compared with the indices obtained from the control area. The difference in yield and expenditure linked with the increase of yield or improvement of the quality of the produce obtained will characterize the effectiveness of presowing irradiation of seeds.

Generally, while conducting the field and commercial trials on presowing irradiation of seeds in a definite climatic zone for a number of years, the collective farms do not retain the control areas and the economic effectiveness of presowing irradiation is determined by estimates with due consideration to the actual expenses on seed irradiation and the data on the increases over several years on the farm or the adjoining research institution with similar agroclimatic conditions.

While determining the economic effectiveness of the presowing irradiation of seeds under conditions of commercial cultivation, it is necessary to consider the expenses of the acquisition of radiation technology, its servicing, as well as the expenses of harvesting and selling the produce.

The general results of the presowing irradiation of seeds in a few regions with one climatic zone make it possible to estimate the expected economic effectiveness of the presowing irradiation of seeds in the remaining collective and state farms in the zone.

Along with the determination of the economic effectiveness of the presowing irradiation of seeds under commercial cultivation conditions, it is also of interest to analyze the use of this new agronomic measure and radiation technology on the scale of the region, province and country (without the intermediate operation of summing up the effect of each crop and for all crop rotations). Such estimates are possible if we consider the groups of major crops in the larger part of the area sown with irradiated seeds.

In several works [193, 194] we find the results of preliminary assessment of the economic effectiveness of the use of experimental and serial commercial models of the Kolos gamma-units with due consideration for the mean values of yield, expenses on production and increased yield, whereby it has been shown that the cost of the irradiation units is recovered during the very first season of their operation. Here are the expenses on the irradiation of seeds for an area of one hectare, in rubles:

Wheat	1.96	Corn for silage making	0.09
Pea	0.84	Sugarbeet	0.06
Corn for grain	0.11		

According to data reported in [220], the average increased corn yield for four years is 4 q/ha (12%) and one hour of operation of the Kolos unit on the farm yields a profit of 1,000–1,300 rubles.

The actual net profit in the Pavlodar province of KazSSR is of great interest [158]. It was 84.5 thousand rubles in 1971 with the use of only one Kolos unit, and 221 thousand rubles in 1972 with the use of two such units and 298 thousand rubles in 1973 when three units were put into use. In 1972, the use of four Kolos units in MSSR yielded a net profit of 547 thousand rubles.

The preliminary assessment of the economic effectiveness of the presowing irradiation of seeds of a group of crops (millet, buckwheat, sunflower, corn) with the commercial gamma-units Kolos has shown that, as a result of additional produce, the economic effect was 6,608 thousand rubles, as compared to the expenditure on the acquisition of 21 Kolos units of only 1,260 thousand rubles (for the conditions of Pavlodar province of KazSSR). An analogous picture may be observed with the wider application of this method in MSSR in the cultivation of corn for grain: as a result of getting 150 thousand tonnes of additional grain output the economic effect is to the tune of nine million rubles, whereas the expenditure on the acquisition of 15–20 Kolos units comes to only about one million rubles.

Recently, methodical instructions [165] have been published, which contain recommendations, as per the norms, to calculate the technical and economic indices of the new agronomic measure and radiation technology.

RADIATION SAFETY IN THE USE OF GAMMA-UNITS

Accumulated data on the use of gamma-units permits us to consider that in the near future, the number of commercial gamma-units such as Kolos, Universal, and Stimulyator will go on increasing and, there will be a simultaneous increase in the number of persons working with them, as well as other people engaged in loading the units with radiation sources and their assembly under conditions such as those at the manufacturing plant.

It is only natural that, as a result of the wide application of radiation sources used in gamma-units in agricultural production, it is necessary to investigate and evaluate all problems concerned with radiation safety.

There are rules of radiation safety and hygiene which are common for different areas of the national economy [61, 159, 160, 174, 177, 209–211]; at the same time some reports [247, 248] contain specific problems of safe working conditions associated with the use of mobile gamma-units in agriculture. The creation of safe working conditions will be facilitated by the production of models of radiation equipment and instrument to measure the ionizing radiation which have been thoroughly tested and investigated in all details before being produced serially by our industry. And also by raising the qualification of the workers engaged in agricultural production and increasing their practical experience by systematic instruction and testing their knowledge.

Any development of a new radiation technology is meaningless without simultaneous solutions to the problems of protection and providing complete safety for the servicing personnel. As a result of many years of investigation, extensive material has been accumulated and generalized which help characterize the overall assessment of the possible harmful effects of radiation and non-radiation factors on various human organs. The obligatory norm of sanitary-hygiene makes it possible to objectively judge if there is any danger to health because of ionizing radiation or the concentration of any other harmful substances in the air and water. It has been established by medical researchers that two types of radiation injuries are possible and may have genetic and somatic after-effects. Genetic after-effects refer to those radiation damages which are transmitted from one generation to the other through structural changes to the genes of the irradiated person. Somatic after-effects mean those changes emerging directly in the person who was subjected to ionizing radiations (one time or chronic exposure) in his life. The dose which will not cause irreversible somatic and genetic changes in the human organism is called the extreme permissible dose (EPD). In the USSR, the concept of the norm of radiation safety conforms to the recognized document [174] and general ideas of the International Committee on Radiation Safety [175].

Assurance of the reliability of hermetization of systems with radiation sources, improved conditions of operation of the sources and the observance of safety rules which includes the timely control over the presence of radioactive pollutions and the use of means of individual protection and so on, make it possible to practically exclude any danger of internal irradiation to personnel at work on modern gamma-units.

Besides the radiation factors, there exist also non-radiation factors that affect a person, such as products of the radiolysis of air—ozone, nitrates, carbon monoxide, ammonia, formaldehyde, hydrocyanic acid [89, 90], toxic poisonous chemicals used for seed dressing [162] and others. Only the assortment of substances produced as a result of the radiolysis of air causes headache, catarrhal inflammation, burning in eyes and throat, insomnia and asthenic condition in humans. Nitrates can combine with blood hemoglobin and transform it into methemoglobin, which cannot carry oxygen. With an ozone concentration of 0.2–1 mg/m^3, we observe in man mucous irritation of the upper respiratory ducts and eyes and headache, vomiting and vertigo with the concentration of 2 mg/m^3 and above (up to 20 mg/m^3).

The existing methods to determine the content of toxic substances [60, 87] enable us to find their concentration even in amounts considerably less than the permissible limits and, accordingly, to ensure the required control over working conditions. The extreme permissible amount of ozone in the air on the premises where the unit is used is 0.1 mg/m^3 and 5 mg/m^3 for nitrates. The threshold of the olfactory perception of ozone is 0.02 mg/m^3 and nitrogen dioxide (the main component of nitric oxide) within a range of 0.05–0.3 mg/m^3. The reports [91,177] contain correlations necessary to calculate the concentrations of ozone, nitric oxides, etc., which develop as a result of ionizing radiation and also the formulas to calculate the corresponding ventilation or air replacement. As the activity of the radiator in commercial gamma-units does not exceed 5–6 curie and good natural ventilation is practically always in the hood of the gamma-unit, it is not required to have a special ventilation mechanism in the gamma-unit.

In the context of compulsory mobility, the radiation sources in the irradiators of the units may be subjected to vibration and shock stresses, which must be reduced to a minimum by perfecting the design of the irradiator and its unit. Designing commercial gamma-units with stationary immobile irradiators and automating and mechanizing the irradiation processes aim to prevent accidents.

At present the commercial gamma-units Kolos, Stimulator and Universal have immobile irradiators and are "self-protected" from the viewpoint of radiation safety; this means that during operation the sources themselves are not mixed and, accordingly, are not subjected to abrasion and mechanical stresses and it is practically impossible to leave the radiation sources open.

For experimental work in certain cases, the use of gamma-units in the research institutions is continued. In working with some units (as a rule, with a mobile irradiator) we require specially equipped premises, whereas with other units (with self protection and as a rule, with an immobile irradiator) specially equipped premises are not required. The first group includes units such as GUBE-800, GUBE-4000, VIESKh and others and the second group refers to gamma-units like Stebel', LMB and Stimulator.

In the context of the industrial production of the Kolos commercial gamma-units, experimental units Stebel, LMB-gamma 1, Issledo Vatel' and others, it became essential to investigate the working conditions and evaluate the actions of radiation and non-radiation factors on service personnel during the operation of the units [155, 247, 248] and while loading of the irradiator. For this purpose the working conditions during the operation of the experimental models and the four serial models of the Kolos units were also studied.

The investigation on the working conditions of the service personnel included an assessment of the field of gamma-radiation and the level of radiation exposure of the service personnel. Measuring the fields of external gamma-radiation on the surface of the container of the unit and inside the hood was done with the RUP-1, SRP-2, MRM or USIT-2 instruments. The strength of the dose at the control desk was 0.1–0.2 mr/hour and it did not exceed 1 mr/hour on the surface of the container, which is considerably below the maximum permissible level in conformity with safety rules No. 774–68. The strength of the dose in the cabinet of the automobile was not more than 0.01–0.02 mr/hour.

Individual doses of irradiation were determined with DK-0.2 and IFK dosimeters which do not have rapid fluctuations in the energy interval from 0.2 to 2 MeV. The results of individual dosimetric control showed that the doses received by the service personnel do not exceed 5–8 mr in a day. The surface pollution of accessible surfaces (regulating disks, internal cone) of joints and parts of the unit was determined to completely verify the radiation by the method of taking moist acidic smears. The test was verified on the experimental model after four years of operation. Smears were taken with a muslin cloth soaked in 10% HNO_3 from smooth surfaces from an area of about 150 cm^2. The radioactive pollution of the smears was determined by the direct determination method on the 100-channel AI-100 analyzer, with an FEU-13 detector and 40 × 40 mm NaI (Tl) crystal. The absolute quantity of ^{137}Cs was determined on the basis of the area of energy peaks; the effectiveness of the estimate was about 9.4%. The backdrop and smear were measured over one hour period. In the context of the given effectiveness of the estimate and the statistically reliable measurement time (reliability 99%), we may find 3.5×10^{-11} curie ^{137}Cs.

No ^{137}Cs was detected in the smears. This testifies to the adequate hermetization of the irradiator of the Kolos unit and the possibility of using this unit under conditions of agricultural production over a prolonged time.

Simultaneously, the quantity of ozone and ozone oxide was also determined in the air inside the hood. Air samples were drawn at the points of feeding and selection of seed material in liter bottle containers, which were evacuated up to 10^{-2} mm of mercury column in advance. After completing the sample collection the analysis was done by the universally accepted method [113, 224]. The concentration of ozone in all cases was less than 2×10^{-5} ml/l, while that of ozone oxide was less than 0.5 ml/l as against the extreme permissible limits of 1.10×10^{-4} and 5×10^{-3} ml/l concentrations respectively.

Taking into consideration the specific design and operation of the gamma-units with immobile irradiator, safety rules (SR No. 774–68) [211] were studied which stipulate certain prerequisites regarding the operation and design of these units. The provisions of SR No. 774–68 made it possible, on the one hand, to concretize the safety provision and, on the other hand, to use the unit more extensively in the interest of national economy. The rules also define specific prerequisites which must necessarily be fulfilled in the process of designing, constructing, loading and operating the powerful isotopic units with immobile irradiators. With the observance of these requirements the units can be used without limitations to conduct different research programs and for the large-scale commercial irradiation of different products.

The rules are applicable to powerful gamma-units with built-in protection shields of hard materials (lead, concrete, pig iron and so on) which do not require specially equipped premises (canyons, chambers and cabinets etc.).

The rules also mention prerequisites for the design and structure of protection shields, protective materials and premises in which gamma-units are housed with an immobile irradiator, as well as for units ready for operation after the installation, loading, boost charging and changing the sources of irradiation. The volume and periods of dosimetric control and the estimate and order of recording the results of measurements are also defined in the rules. Requirements for prophylactic inspections and prophylactic repairs are also given. A special chapter is devoted to measures to prevent accidents.

On the whole, safety rule No. 774–68 has made it possible to reduce many prerequisites stipulated previously, vide SR No. 333–60* and SR No. 482–64. For example, prerequisites pertaining to housing the units, air replacement, conditions of operation, dosimetry on the units and so on, which had facilitated not only the provision of radiation safety in the use of the units, but also their even more extensive use in agriculture.

The irradiator of experimental gamma-units relates to the first group and has a total activity up to 6.5 Kc. Depending on the protection shield from ionizing radiations, these units can be subdivided into:

units with "dry" protection, in which the irradiation chamber and store of the radiation sources are made of hard materials such as concrete, steel, pig iron and others;

units with "wet" protection, in which the storage of the radiation sources and the irradiation chamber is placed under a layer of liquid such as water, oil and others. This type of protection is often called "watery" because water is used as the protection shield material.

units with mixed protection, in which the storage of the radiation sources is done under water, but irradiation is done in the work chamber made of hard materials, where the source is placed from the store at the time of irradiation.

In our country, the first gamma-units to conduct research on the presowing irradiation of seeds, viz., GUPOS (Stebel') was built in 1957–1958 and self-protection units practically replaced the units of the second group.

In conclusion, we will examine the problems of general control of measures concerning the provisions of radiation safety, including dosimetric and

*At present on the basis of the requirements of NRB-69 and recommendation of the MKRZ a new normative document had been confirmed, viz., "Basic sanitary rules of working with radioactive substances and other sources of ionizing radiation (OSP-72)" [176].

radiometric control, which must be performed by the service units for radiation safety [209, 211, 252] under the conditions of agricultural production and in research institutions. In the absence of the service of radiation safety units or personnel, its functions may be assigned to a specialist with corresponding qualifications for servicing the unit or a group of units. The radiation safety service exercises control over individual doses of radiation, level of external background radiation at working places, effectiveness of protection against radiation, radioactive pollution of working premises and equipment and the proper functioning of the systems of blocking and signal monitoring. The results of dosimetric control must be recorded in a special register. On the basis of the data obtained, the integral doses of irradiation of the service personnel are calculated every week, month and year of work. Individual cards are maintained for each person working on the unit, which shows the monthly and yearly doses of external background radiation, as well as data about the level of pollution of surfaces in the work place and the participation of individuals in repairs connected with radiation danger.

The following periodicity of control is recommended:

individual dosimetric control—every day;

measurement of the strength of dose at working places—not less than once in a month;

detailed determination of the strength of dose on the surface of protection shield from radiation during the first year of operation of the unit—not less than once a month, but afterward—not less than twice a year;

measurement of radioactive pollution in gamma-units with wet or mixed protection—not less than once in a month;

control for the proper functioning of the systems of blocking and signal monitors—daily.

The periodicity of prophylaxis and conducting the prophylactic repair works is determined by the administrative institution (enterprise) and designer-organization.

All persons who service the gamma-units must undergo prior medical examination and should be allowed to start work only in the absence of any positive medical indications as envisaged by the safety rules.

Personnel engaged in the operation of units of any type must know the purpose of each adjunct and the unit as a whole. They must study its design and servicing system, as well as correct and safety work procedures during prophylactic repairs.

During the operation of experimental and commercial gamma-units for the presowing irradiation of seeds, it is necessary to have instruments to determine the dose strength on the surface of protection shield and to conduct individual checks or controls.

The characteristics of instruments to be used for individual checking are given below:

Table 5.11. Some characteristics of the UIM2-eM instrument

Detection block	Range of measurements, r/hour	Speed of count, counts/sec
DG-1A*	10^{-4}–10^{-1}	1.7–1.5×10^{3}
DG-1B**	10^{-4}–10^{-1}	1.7–1.5×10^{3}
DG-2A	10^{-3}–1	3.0–2.5×10^{3}
DG-2B	10^{-3}–1	3.0–2.5×10^{3}
DG-3A	10^{-2}–50	2.6–9.0×10^{3}
DG-3B	10^{-2}–50	2.6–9.0×10^{3}
DG-7A	0.1–300	5.7–1.2×10^{4}
DG-7B	0.1–300	5.7–1.2×10^{4}
DG-8A	10^{-4}–10^{-1}	1.7–1 5×10^{3}
DG-9A	10^{-4}–10^{-1}	1.7–1.5×10^{3}
DG-9B	0.1–300	5.7–1.2×10^{4}
DG-10A	10^{-6}–1	3–2.500
DG-10B	0.1–330	5.7–1.2×10^{4}
DG-11A	11^{-2}–50	2.6–9.0×10^{3}
DG-11B	3.1–300	5.7–1.2×10^{4}

*Aluminum body.
**Stainless steel body.

A set of individual IFK-2.3 photofilm dosimeters. Range of measurements of gamma-radiation 0.01–50 r, with the energy 0.02–3.0 MeV, error of measurement of dose of irradiation in a range of ± 20%.

A set of individual IFKU photofilm dosimeters. Range of measurement of gamma-radiation 0.05–2 rem, with energy 0.1–3 MeV, error of measurement of the dose ± 10%.

A set of individual DK-02 or KID dosimeters.

Individual DUG 2-01 signal dosimeters which have the following technical features:

The range of measurement of exposure dose of gamma-radiation of energy from 80 KeV to 2 MeV, mr	0–500 (1st reference point 25 mr)
Graduation of the indicator of strength of operational dose of ^{60}Co gamma-radiation point-wise:	
with the meter SBM-10	50 mc/hour,
Sound indication	300 mc/hour, 1 mr/hour
light indication	10 mr/hour, 100 mr/hour
with the meter SI-34 G:	
Sound indication, mr/hour	1, 10, 50
light indication, r/hour	1, 10, 50

(Contd.)

Error of measurement of the dose of irradiation, % of the complete scale:	
in the range of energies 0.2–2 MeV	not more than ±10
in the range of energies 80–220 KeV	not more than ±30
Self-discharge of dosimeter, % of the maximum value, of scale:	
for 8 hours of work at 0–40°C	not more than ±10
Overall size, mm	25 × 64 × 126

For the dosimetric checking of gamma-fields and effectiveness of protection shields, a UIM2-eM instrument (which replaces USIT-2) is being produced, which, even in working with detection blocks, enables us to measure gamma-radiation with an error of ±10% (Table 5.11).

The SGD-1 instrument (Kura) can also be used instead of SRP-2 or MRM-2.

In the near future, commercial gamma-units will enrich the machine-tractor parks of our collective and state farms. Their wider application will be facilitated by the operation of radiation technology, satisfying all requirements of safety materials [174, 176], with due consideration for the specific use of the units under practical conditions of agricultural production.

Bibliography

1. Abdulaev, M.A. 1970. Vliyanie gamma-oblucheniya semyan na urozhainost' i nekotorye biokhimicheskie protsessy razlichnykh ovoshchnykh kul'tur (The effect of gamma-irradiation on yield and some biochemical processes of different vegetable crops). Dissertation for Ph.D. degree (Candidate of biological sciences) (Institute of Botany named after V.L. Komarov).
2. Abutalybov, M.G. and N.B. Vezirova. 1958. Deistvie radiatsii na okislitel'no-vosstanovitel'nye protsessy v rastitel'nom organizme (Effect of radiation on oxidation-reduction processes in the plant organism). In the book: *Trudy Vsesoyuznoi Konferentsii Po Primeneniyu Radioaktivnykh i Stabil'nykh Izotopov v Narodnom Khozyaistve i Nauke (4–12 April, 1957)*. Izd-vo AN SSSR, Moscow, pp. 138–143.
3. Agakii, P.G. and K.I. Sukach. 1972. Vliyanie predposevnogo oblucheniya semyan na produktivnost' sakharnoi svekly (Effect of pre-sowing irradiation of seeds on the productivity of sugarbeet). In the book: *Tezisy Dokladov Vsesoyuznoi Konferentsii Po Ispol'zovaniyu Radiatsionnoi Tekhniki v Sel'skom Khozyaistve*, Kishinev, vol. 1, pp. 42–43.
4. Agakishev, D.A. 1958. Vliyanie radioaktivnykh izotopov na razvitie khlopchatnika (Effect of radioactive isotopes on the development of cotton). *Izv. AN TSSR*, No. 6, pp. 99–103.
5. Aglintsev, K.K. 1957. *Dozimetriya Ioniziruyushchikh Izluchenii* (Dosimetry of ionizing radiation). Second edition. Gostekhizdat, Moscow.
6. Aleksandrova, I.F. 1970. Metabolizm i produktivnost' teplichnykh tomatov v svyazi s gamma-oblucheniem semyan (Metabolism and productivity of glass-house tomatoes in the context of gamma-irradiated seeds). Dissertation for Ph.D. degree (Candidate of biological sciences) (Gorky State University).
7. Allen, A.O. 1963. *Radiatsionnaya Khimiya Vody i Vodnykh Rastvorov* (Radiation chemistry of water and water solutions). Translation from English, Gosatomizdat, Moscow.
8. Andreev, V.S. 1963. Geneticheskii mekhanizm radio-stimulyatsii rastenii (Genetic mechanism of radio plant stimulation). In the book: *Predposevnoe Obluchenie Semyan Sel'skokhozyaistvennykh Kul'tur*. Izd-vo AN SSSR, Moscow, pp. 28–38.
9. Antoshkevich, V.S. 1967. *Ekonomicheskaya Effektivnost' Sel'skokhozyaistvennykh Mashin* (Economic effectiveness of agricultural machines). "Ekonomika" Publishers, Moscow.

10. Arifov, U.A., G.A. Klein and S.A. Anastasov. 1963. Deistvie predposevnogo oblucheniya semyan na rost i napravlennoe izmenenie svoistv nekotorykh sel'skokhozyaistvennykh kul'tur Uzbekistana (The effect of presowing irradiation of seeds on the growth and induced change of properties of some agricultural crops of Uzbekistan). In the book: *Predposevnoe Obluchenie Semyan Sel'skokhozyaistvennykh Kul'tur*. Izd-vo AN SSSR, Moscow, pp. 164–169.

11. Atabekova, A.I. 1936. Dei stvie rentgenovskikh luchei na semena i prorostki gorokha (Effect of x-rays on pea seeds and seedling). *Biol. Zhurnal*, vol. 5, No. 1, pp. 99–116.

12. Atabekova, A.I. 1936. Deistvie rentgenovskikh luchei na pokoyashchiesya i prorastayushchie semena (Effect of x-rays on dormant and germinating seeds). *Biol. Zhurnal*, vol. 5, No. 25, pp. 234–240.

13. Atabekova, A.I. 1937. O vliyanii luchei Rentgena na rost i razvitie gorokha (Effect of x-rays on the growth and development of pea). *Biol. Zhurnal*, vol. 6, No. 1, pp. 81–92.

14. Atabekova, A.I. 1938. O stimuliruyushchem deistvii luchei Rentgena na rasteniya (Stimulation effect of x-rays on plants). *Priroda*, No. 78, pp. 56–62.

15. Afanas'eva, A.S. 1936. Sokhranyaemost' deistviya rentgenovskikh luchei na pshenitsu (Durability of the effect of x-rays on wheat). *Byul. Moskovsk. O-va Ispytatelei Prirody. Otd-nie Biol.*, vol. 45, pp. 433–440.

16. Afanas'eva, A.S. 1938. Deistvie x-luchei na delenie yadra v konchikakh kornei pshenitsy (Effect of x-rays on the division of the nucleus in the tips of wheat roots). *Biol. Zhurnal*, vol. 7, No. 2, pp. 189–194.

17. Afanas'eva, A.S. 1939. Deistvie rentgenovskikh luchei na vozdushnosukhie, uvlazhnennye i prorosshie zerna pshenitsy (Effect of x-rays on air dried, soaked and germinated wheat grain). *Byul. Moskovsk. O-va Ispytatelei Prirody. Otd-nie Biol.*, vol. 48, pp. 19–28.

18. Afanas'eva, A.S. 1941. Sravnitel'noe izuchenie mikrosporogeneza u pshenits (normal'noi i obluchennoi rentgenovskimi luchami) [Comparative study of wheat microsporogenesis (normal and irradiated by x-rays)]. *Izv. AN SSSR, Ser.-Biol.* No. 2, pp. 224–243.

19. Batov, V.V. and Yu.I. Koryakin. 1969. *Ekonomika Yadernoi Energetiki* (Economics of Nuclear Energetics). Atomizdat, Moscow.

20. Batygin, N.F. 1963. K Voprosu o ponimanii protsessov radio-stimulyatsii (The problem of the concept of radio-stimulation processes). In the book: *Predposevnoe Obluchenei Semyan Sel'skokhozyaistvennykh Kul'tur*. Izd-vo AN SSSR, Moscow, pp. 21–27.

21. Berezina, N.M. 1964. Predposevnoe Obluchenie Semyan Sel'skokhozyaistvennykh Rastenii (Presowing irradiation of the seeds of agricultural plants). Atomizdat, Moscow.

22. Berezina, N.M. 1960. Primenenie ioniziruyushchikh izluchenii dlya povysheniya urozhainosti sel'skokhozyaistvennykh kul'tur (Use of ionizing radiation to increase the yield of agricultural crops). *Atomnaya Energiya*, vol. 9, No. 5, pp. 432–437.
23. Berezina, N.M. 1963. Rol' vneshnikh faktorov i fiziologicheskogo sostoyaniya semyan v effekte stimulyatsii (The role of external factors and the physiological condition of seeds on the stimulation effect). In the book: *Seminar Po Primeneniyu Istochnikov Ydernykh Izluchenii Dlya Povysheniya Urozhainosti Sel'skokhozyaistvennykh Kul'tur i Polucheniya Novykh Khozyai stvenno-tsennykh Form Rastenii.* (VDNKh) Moscow, pp. 3–4.
24. Berezina, N.M. 1964. Ispol'zovanie predposevnogo oblucheniya semyan kukuruzy dlya povysheniya urozhaya i uluchsheniya kachestva syr'ya (The use of presowing irradiation of corn seeds to increase the yield and improve the quality of raw produce). In the book: *Biologicheskie Osnovy Povysheniya Kachestva Semyan Sel'skokhozyaistvennykh Rastenii.* "Nauka" Publishers, Moscow, pp. 167–171.
25. Berezina, N.M. 1970. Teoreticheskie osnovy predposevnogo oblucheniya semyan sel'skokhozyaistvennykh rastenii i puti ego prakticheskogo ispol'zovaniya (Theoretical bases of the presowing irradiation of agricultural plant seeds and its practical application). In the book: *Materialy Pervoi Nauchno-prakticheskoi Konferentsii Po Primeneniya Izotopov i Ioniziyruyushchikh Izluchenii v Sel'skom Khozyaistve.* Kishinev, pp. 5–6.
26. Berezina, N.M. 1972. Uroven' sovremennykh issledovanii po predposevnomu oblucheniyu semyan sel'skokhozyaistvennykh rastenii (The level of contemporary research on presowing irradiation of agricultural plant seeds). In the book: *Tezisy Dokladov Vsesoyuznoi Konferentsii Po Ispol'zovaniyu Radiatsionnoi Tekhniki v Sel'skom Khozyaistve*, Kishinev, vol. 1, pp. 7–8.
27. Berezina, N.M. 1966. Stimuliruyushchee deistvie gamma-luchei pri predposevnom obluchenii semyan sel'sko khozyaistvennykh rastenii i nekotorye faktory, ego obuslovlivayushchie (Effect of gamma-rays in the presowing irradiation of agricultural plant seeds and some factors stimulating it). In the book: *Ioniziruyushchie Izlucheniya v Rastenievodstve*, "Sovetskaya Kuban' " Publishers, Krasnodar, pp. 6–10.
28. Berezina, N.M. 1971. Sovremennoe sostoyanie voprosa o predposevnom obluchenii semyan sel'skokhozyaistvennykh kul'tur u nas i za rubezhom (Contemporary status of the problem of presowing irradiation of seeds of agricultural crops in our country and abroad). In the book: *Deistvie Radiatsii Na Rasteniya*, "Fan" Publishers,Ta shkent, pp. 4–17.
29. Berezina, N.M. 1971. Primenenie metoda predposevnogo oblucheniya semyan gamma-kvantami dlya povysheniya urozhainosti sel'skokho-

zyaistvennykh kul'tur (Use of the method of presowing irradiation of seeds by gamma-radiation to increase the yield of agricultural crops). *Izotopy v SSSR*, No. 2, pp. 8–10.

30. Berezina, N.M. and M.A. Abdulaev. 1967. Uluchshenie kachestva tomatov v protsesse sozrevaniya (Improvement of tomato quality in the ripening process). In the book: *Zapiski Gosudarstvennogo Universiteta Az SSR*. Baku. pp. 35–38.
31. Berezina, N.M. and M.A. Abdulaev. 1968. Radiostimulyatsionnyi effekt pri predposevnom obluchenii semyan tomatov (Radio stimulation effects during the presowing irradiation of tomato seeds). *Doklady AN AzSSR*, vol. 24, No. 1, pp. 38–41.
32. Berezina, N.M. and M.A. Abdulaev. 1969. Vliyanie predposevnogo gamma-oblucheniya semyan salata, ukropa u redisa na kachestvo urozhaya (Effect of presowing gamma-irradiation of the seeds of lettuce, dill and radish on the quality of the produce). *Vestn. S. Kh. Nauki AzSSR*, No. 6, pp. 18–22.
33. Berezina, N.M. and T.E. Guseva. 1968. Deistvie ioniziruyushchikh izluchenii na nekotorye maslichnye kul'tury (Effect of ionizing radiation on some oilseed crops). In the book: *Vsesoyuznaya Nauchnaya Konferentsiya Po Primeneniyu Izotopov i Izluchenii v Sel'skom Khozyaistve. 20–24 iyunya 1967 g*. Moskva. pp. 25–26 (VASKhNIL).
34. Berezina, N.M. and T.E. Guseva. 1970. Razlichnaya radiochuvstvitel'nost' rastenii podsolnechnika pri khronicheskom obluchenii ikh gammaluchami ^{60}Co v raznykh fazakh razvitiya (Different radio sensitivities of sunflower plants under chronic irradiation by ^{60}Co gamma-rays at different developmental phases). *Radiobiologiya*, vol. 10, No. 2, p. 313.
35. Berezina, N.M., E.I. Korneva and R.R. Riza-Zade. 1962. Itogi proizvodstvennogo ispytaniya priema predposevnogo oblucheniya semyan kukuruzy gamma-luchami ^{137}Cs (Results of large-scale, commercial trials of the presowing irradiation of corn seeds by ^{137}Cs gamma-rays). *Radiobiologiya*, vol. 2, No. 4, pp. 629–633.
36. Berezina, N.M. and A.A. Narimanov. 1969. Reparatsionnye protsessy pri postradiatsionnom progreve semyan khlopchatnika, obluchennykh v doze 40 c (Repair processes during post-irradiation heating of cotton seeds irradiated with the dose of 40 c). *Radiobiologiya*, vol. 1, No. 2, pp. 262–265.
37. Berezina, N.M., A.F. Revin and A.A. Narimanov. 1968. Izmenenie soderzhaniya triptofana v nekotorykh sel'skokhozyaistvennykh kul'turakh pod vliyaniem oblucheniya (Change of tryptophane content in some agricultural crops due to irradiation). In the book: *Vsesoyuznaya Nauchnaya Konferentsiya Po Primeneniyu Izotopov i Izluchenii v Sel'skom Khozyaistve*. p. 11, VASKhNIL, Moscow.
38. Berezina, N.M. and R.R. Riza-Zade. 1963. Proizvodstvennaya pro-

verka priema predposevnogo oblucheniya markovi i kapusty (Commercial trial of the presowing irradiation of carrot and cabbage). In the book: *Predposevnoe Obluchenie Sel'skokhozyaistvennykh Kul'tur*. Izd-vo AN SSSR, Moscow, pp. 194–198.

39. Berezina, N.M. and others. 1960. Primenenie gamma-oblucheniya pri vyrashchivanii zemlyaniki (Use of gamma-irradiation in strawberry cultivation). *Pishchevaya Promyshlennost'* No. 1(6), pp. 26–28.

40. Berezina, N.M. and others. 1961. Povyshenie urozhainosti i vitaminnosti zemlyaniki pri vyrashchivanii posadochnogo materiala na gammapole (Increase of strawberry yield and vitamin content while growing the planting material in gamma fields). In the book: *Vitaminy*. Pishchepromizdat Publishers, Moscow, vol. 8, pp. 86–92.

41. Berezina, N.M. and others. 1971. Issledovanie otvetnoi reaktsii rastenii na razlichnye fizicheskie vozdeistviya (Study of reciprocal reaction of plants to different physical actions). In the book: *Deistvie Radiatsii Na Rasteniya*, "Fan" Publishers, Tashkent, pp. 32–34.

42. Berezina, N.M. and others. 1963. Vliyanie predposadochnogo oblucheniya Klubnei gamma-luchami ^{60}Co na urozhai i soderzhanie vitamina C v kartofele (Effect of preplanting irradiation by ^{60}Co gammarays on the yield and vitamin C content of potato tubers). *Radiobiologiya*, vol. 3, No. 1, pp. 139–143.

43. Berezina, N.M. and others. 1968. Vosstanovlenie radiatsionnykh povrezhdenii semyan khlopchatnika v protsesse dlitel'nogo Khraneniya (Healing radiation injuries of cotton seeds during prolonged storage). *Radiobiologiya*, vol. 8, No. 3, pp. 438–441.

44. Bibergal', A.V., V.I. Sinitsin and N.I. Leshchinskii. 1960. *Izotopnye Gamma-Ustanovki* (Isotope gamma-units). Atomizdat, Moscow.

45. Bibergal', A.V. and others. 1962. Transpartabel'naya gamma-ustanovka GUPOS-^{137}Cs-800 dlya predposevnogo oblucheniya semyan (Mobile gamma-unit GUPOS-^{137}Cs-800 for the presowing irradiation of seeds). *Atomnaya Energiya*, vol. 12, No. 2, pp. 159–160.

46. Bibergal', A.V. and others. 1958. Gamma-ustanovka GUBE-800 dlya radiobiologicheskogo eksperimenta (Gamma-unit GUBE-800 for radiobiological experiments). *Biofizika*, vol. 3, No. 1, pp. 118–121.

47. Bordyuzhevich, V.G., K.I. Sukach and R.I. Cheban. 1972. Vliyanie predposevnogo oblucheniya semyan na rost, razvitie, urozhai i kachestvo podsolnechnika (Effect of presowing irradiation on the growth, development, yield and quality of sunflower seeds). In the book: *Tezisy Dokladov Vsesoyuznoi Konferentsii Po Ispol'zoyaniyu Radiatsionnoi Tekhniki v Sel'skom Khozyaistve*. Kishinev, vol. 1, pp. 122–123.

48. Borisevich, V.A. and others. 1972. Opyt ispol'zovaniya moshchnoi gamma-utsanovki UGU-200 Instituta yadernoi energetiki AN BSSR dlya obrabotki sel'skokhozyaistvennoi produktsii (The experiment of

using the powerful gamma-unit UGU-200 from the Institute of Nuclear Energetics of the Academy of Sciences of BSSR to treat agricultural produce). In the book: *Tezisy Dokladov Vsesoyuznoi Konferentsii Po Ispol'zovaniyu Radiatsionnoi Tekhniki v Sel'skom Khozyaistve*, Kishinev, vol. 3, pp. 6–7.

49. Borodin, P.R., G.L. Korikova and V.M. Lebezhennikova. Nekotorye rezul'taty vozdeistviya ioniziruyushchikh izluchenii na kukuruzu (Some results of ionizing radiation on corn). In the book: *Trudy Sverdlovskogo Sel'skokhozyaistvennogo Instituta*. Sverdlovsk, vol. 7, pp. 239–244.
50. Bregadze, Yu.I. and others. 1970. Eksperimental'nye metody issledovaniya doznykh polei v rabochikh kamerakh seriinykh izotopnykh gamma-ustanovok (Experimental methods for the study of dose fields in the working chambers of serial isotope gamma-units). In the book: *Materialy Pervoi Nauchnoprakticheskoi Konferentsii Po Primeneniyu Izotopov i Ioniziruyushchikh Izluchenii v Sel'skom Khozyaistve*. Kishinev, p. 53.
51. Breslavets, L.P. 1935. Rezul'taty rabot po vozdeistiyu rentgenovskikh luchei na semena sel'skokhozyaistvennykh rastenii (Results of research on the effect of x-rays on the seeds of agricultural plants). *Elektrifikatsiya Sel'skogo Khozyaistva*, vol. 2, pp. 38–42.
52. Breslavets, L.P. 1937. Sovremennoe sostoyanie rentgenologii rastenii (Contemporary status of plant radiology). *Byul. Moskovsk. O-va Ispytatelei Prirody. Otd-nie Biol.*, vol. 46, pp. 359–369.
53. Breslavets, L.P. 1940. Izmeneniya, vyzvannye v rasteniyakh luchami Rentgena i znachenie etikh izmenenii dlya teorii i praktiki (Changes caused by x-rays in plants and the theoretical and practical importance of these changes). *Trudy Botanicheskogo Sada MGU*, No. 3, pp. 75–81.
54. Breslavets, L.P. 1946. *Rasteniya i Luchi Rentgena* (Plants and X-rays). Izd-vo AN SSSR, Moscow-Leningrad.
55. Breslavets, L.P. 1956. Deistvie ioniziruyushchikh izluchenii na rost i razvitie nekotorykh sel'skokhozyaistvennykh rastenii (Effect of ionizing radiation on the growth and development of some agricultural plants). *Biofizika*, vol. 1, No. 7, pp. 628–631.
56. Breslavets, L.P. and A.I. Atabekova. 1935. Povyshenie urozhaya pod vliyaniem rentgenovskikh luchei (Increase of yield due to the effects of x-rays). In the book: *Trudy Vsesoyuznogo Nauchno-Issledovatel'skogo Instituta Udobrenii i Agropochvovedeniya*. Sel'khozgiz, Moscow, vol. 8, pp. 254–258.
57. Breslavets, L.P. and A.S. Afanas'eva. 1935. Deistvie rentgenovskikh luchei na rozh' (Effect of x-rays on rye). *Vestn. Rentgenologii i Radiologii*, vol. 14, pp. 288–301.
58. Breslavets, L.P., N.M. Berezina and G.I. Shchibrya. 1956. Dlitel'noe deistvie malykh doz gamma-luchei na nekotorye sel'skokhozyaistven-

naye rasteniya (Long-term effect of low doses of gamma-rays on some agricultural plants). *Biofizika*, vol. 1, No. 6, pp. 555–559.

59. Breslavets, L.P., G.B. Medvedeva and A.S. Afanas'eva. 1935. Povyshenie urozhainosti pod vliyaniem rentgenovskikh luchei (Increase of yield due to the effects of x-rays). In the book: *Trudy Vsesoyuznogo Nauchno-Issledovatel'skogo Instituta Udobrenii i Agropochvovedeniya.* Sel'khozgiz, Moscow, vol. 8, pp. 237–242.

60. Bykhovskaya, M.S. and others. 1966. *Metody Issledovaniya Vrednykh Veshchestv v Vozdukhe v Proizvodstvennykh Pomeshcheniyakh* (Methods to study harmful substances in the air at the production premises). "Meditsina" Publishers, Moscow.

61. Bykhovskii, A.V., A.V. Larichev and E.D. Chistov. 1970. *Voprosy Zashchity Ot Ioniziruyushchikh Izluchenii v Radiatsionnoi Khimii* (The problems of protection against ionizing radiation in radiation chemistry). Atomizdat, Moscow.

62. Vezirova, N.B. 1961. Vliyanie radioaktivnykh izotopov i yadernykh izluchenii na rost, razvitie i obmen veshchestv rastenii (Effect of radioactive isotopes and nuclear radiation on the growth, development and metabolism of plants). In the book: *Trudy Azerbaidzhanskogo Nauchno-issledovatel'skogo Instituta Zemledeliya*, AzNIIZ, Baku, pp. 35–40.

63. Vikulin, A.A. and others. 1969. Malogabaritnaya ustanovka dlya oblucheniya (A small unit for irradiation). *Atomnaya Energiya*, vol. 27, No. 2, p. 174.

64. Vlasov, N.S. 1968. *Metodika Ekonomicheskoi Otsenki Sel'skokhozyaistvennoi Tekhniki* (The method for the economic evaluation of agricultural technology). "Kolos" Publishers, Moscow.

65. Vlasyuk, P.A. 1955. Deistvie yadernykh izluchenii na rasteniya (The effect of nuclear radiation on plants). In the book: *Sessiya AN SSSR Po Mirnomu Ispol'zovaniyu Atomnoi Energii. 1-5 Iyulya 1955.* Izd-vo AN SSSR, Moscow.

66. Vlasyuk, P.A. 1955. Vliyanie malykh doz ioniziruyushchikh izluchenii na sel'skokhozyaistvennye rasteniya (The effects of low doses of ionizing radiation on agricultural plants). In the book: *Trudy Sessii, Posvyashchennoi Dostizheniyam i Zadacham Sovetskoi Biofiziki v Sel'skom Khozyaistve.* Izd-vo AN SSSR, Moscow, pp. 43–52.

67. Vlasyuk, P.A. 1956. *Mikroelementy i Radioaktivnye Izotopy v Pitanii Rastenii* (Microelements and radioactive isotopes in plant nutrition). Izd-vo AN Ukr SSR, Kiev.

68. Vlasyuk, P.A. 1957. Osnovnye zakonomernosti biologicheskogo deistviya malykh doz yadernykh izluchenii (Basic laws of the biological effect of low doses of nuclear radiation). *Doklady VASKhNIL*, vol. 10, pp. 8–15.

69. Vlasyuk, P.A. 1959. Vliyanie malykh doz ioniziruyushchikh izluchenii

na rasteniya (The effect of low doses of ionizing radiation on plants). In the book: *Rost Rastenii*. L'vov state university, L'vov, pp. 363–369.

70. Vlasyuk, P.A. and N.I. Vidzilya. 1958. Vliyanie gamma-chastits radioaktivnykh izotopov na izmenenie velichiny khloroplastov (The effect of gamma-radiation of radioactive isotopes on the change of chloroplast size. *Doklady AN SSSR*, vol. 119, No. 1, pp. 65–67.
71. Vlasyuk, P.A. and D.M. Grodzinskii. 1958. Ob otvetnykh reaktsiyakh rasteniya na yadernykh izlucheniya (Reciprocal reactions of plants to nuclear radiation). In the book: *Trudy Instituta Fiziologii Rastenii AN Ukr SSR*, Izd-vo AN Ukr SSR Kiev, vol. 15, pp. 44–49.
72. Vlasyuk, P.A. and D.M. Grodzinskii. 1956. K voprosu o deistvii yadernykh izluchenii na rasteniya (The problem of the effect of nuclear radiation on plants). *Doklady AN SSSR*, vol. 106, No. 3, pp. 562–566.
73. Vlasyuk, P.A., Z.M. Klimovitskaya and E.S. Kosmatyi. 1956. Deistvie malykh doz ioniziruyushchikh izluchenii na okislitel'no-vosstanovitel'nye protsessy v rasteniyakh (The effect of low doses of ionizing radiation on the oxidation-reduction processes in plants). *Doklady AN SSSR*, vol. 106, No. 4, pp. 731–736.
74. Vlasyuk, P.A., A.V. Manorik and D.M. Grodzinskii. 1963. Vliyanie predposevnoi obrabotki semyan ioniziruyushchei radiatsiei na produktivnost' rastenii (The effect of the presowing treatment of seeds with ionizing radiation on the productivity of plants). In the book: *Predposevnoe Obluchenie Semyan Sel'skokhozyaistvennykh Kul'tur*. Izd-vo AN SSSR, Moscow, pp. 42–53.
75. Voropaev, Yu.V., V.S. Lur'e and R.A. Srapen'yants. 1966. Primenenie gamma-izlucheniya dlya obrabotki kartofelya pered posadkoi (The use of gamma-radiation to treat potatoes before planting). *Izotopy v SSSR*, No. 4, pp. 24–26.
76. Generalova, V.V., M.N. Gurskii and D.A. Kaushanskii. 1972. Standardizatsiya dozimetricheskikh izmerenii na izotopnykh gamma-ustanovkakh s nepodvizhnym obluchatelem (Standardization of dosimetric measurements on isotope gamma-units with an immobile irradiator). In the book: *Tezisy Dokladov Vsesoyuznoi Konferentsii Po Ispol'zovaniyu Radiatsionnoi Tekhniki v Sel'skom Khozyaistve*. Kishinev, vol. 3, p. 34.
77. Gol'dberg, S.V. 1904. K ucheniyu o fiziologicheskom deistvii bekkerelevskikh luchei (The concept about physiological action of Becquerel rays). Thesis for a Doctor of Biological Sciences degree. SPb.
78. Grebinskii, S.O. and V.G. Tsibukh. 1963. Vliyanie rentgenovskogo oblucheniya naklyunuvshikhsya semyan sakharnoi, kormovoi svekly i ovoshchnykh kul'tur na urozhai (The effects of x-radiation on the yield of sprouted seed of sugar and fodder beet and vegetable crop). In the book: *Predposevnoe Obluchenie Semyan Sel'skokhozyaistvennykh Kul'tur*. Izd-vo AN SSSR, Moscow, pp. 179–183.

79. Grechushnikov, A.I. and V.S. Serebrenikov. 1963. Vliyanie predposadochnogo Oblucheniya Klubnei Kartofelya gamma-luchami na rost i razvitie rastenii, urozhai i Kachestvo Kartofelya (The effects of preplanting irradiation of potato tubers by gamma-rays on their growth, development, yield and quality). In the book: *Predposevnoe Obluchenie Semyan Sel'skokhozyaistvennykh Kul'tur*. Izd-vo AN SSSR, Moscow, pp. 94–106.
80. Guseva, V.A. 1963. Vliyanie malykh doz gamma-luchei ^{60}Co na biokhimizm i urozhai rastenii (The effects of low doses of ^{60}Co gamma-rays on the biochemistry and yield of plants). In the book: *Predposevnoe Obluchenie Semyan Sel'skokhozyaistvennykh Kul'tur*. Izd-vo AN SSSR, Moscow, pp. 72–78.
81. Guseva, V.A., L.N. Kurganova and G.M. Gorlikova. 1971. Nekotorye pokazateli energeticheskogo urovnya prorostkov grechikhi v svyazi s deistviem ioniziruyushchei radiatsii na semena (Some indices of the energetic level of rye seedlings in the context of the effect of ionizing radiation on seeds). In the book: *Deistvie Radiatsii Na Rasteniya*. "Fan" Publishers, Tashkent, pp. 34–35.
82. Guseva, V.A. and others. 1970. Predposevnoe obluchenie semyan v usloviyakh Gor'kovskoi oblasti (Presowing irradiation of seeds under the conditions in Gorky province). In the book: *Materialy Pervoi Nauchno-prakticheskoi Konferentsii Po Primeneniyu Izotopov i Ioniziruyushchikh Izluchenii v Sel'skom Khozyaistve*. Kishinev, pp. 31–32.
83. Guseva, T.E. 1969. Razlichnaya radiochuvstvitel'nost' nekotorykh maslichnykh Kul'tur i faktory obuslovlivayushchie ee izmeneniya (Different radio-sensitivities of some oilseed crops and factors causing these changes). Thesis for the Ph.D. degree of the Candidate of Biological Sciences, Institute of Biophysics of the USSR Academy of Sciences, Moscow.
84. Dedul', F.A. 1967. Vliyanie predposevnogo oblucheniya semyan ioniziruyushchei radiatsiei na urozhai ozimoi pshenitsy (The effects of the presowing irradiation of seeds by ionizing radiation on the yield of winter wheat). In the book: *Trudy Nauchno-issledovat'skogo Instituta Zemeledeliya Minsel'khoza GSSR*. Tbilisi, vol. 14, pp. 107–113.
85. Dedul', F.A. and others. 1970. Vliyanie Srokov Khraneniya gamma-obluchennykh semyan pered posevom na nekotorye fiziologicheskie i biokhimicheskie pokazateli i urozhai ozimoi pshenitsy (Effects of the length of storage of gamma-irradiated seeds before sowing on some physiological and biochemical indices and yield of winter wheat). In the book: *Materialy Pervoi Nauchno-prakticheskoi Konferentsii Po Primeneniyu Izotopov i Ioniziruyushchikh Izluchenii v Sel'skom Khozyaistve*. Kishinev, pp. 36–38.
86. Dedul', F.A. and others. 1970. Deistvie stimuliziruyushchikh doz i

moshchnostei doz gamma-luchei na urozhai i biokhimicheskii sostav zerna kukuruzy (The effect of stimulation doses and the strength of the gamma-ray doses on the yield and biochemical composition of corn). In the book: *Materialy Pervoi Nauchno-prakticheskoi Konferentsii Po Primeneniyu Izotopov i Ioniziruyushchikh Izluchenii v Sel'skom Khozyaistve*. Kishinev, pp. 60–62.

87. Demidov, A.V. and L.A. Mokhov. 1962. *Uskorenie Metoda Opredeleniya v Vozdukhe Vrednykh Gazoobraznykh i Paroobraznykh Veshchestv* (Express method to determine harmful gaseous and vapor substances in the air). Medgiz, Moscow.

88. Dzhelepov, B.S. and L.K. Peker. 1958. *Skhemy Raspada Radioaktivnykh Yader* (Disintegration schemes of radioactive nuclei). Izd-vo AN SSSR, Moscow-Leningrad.

89. Dmitriev, M.T. 1964. Dozy ioniziruyushchego izlucheniya, vliyayushchie na sostav atmosfery radiatsionnykh laboratorii (The doses of ionizing radiation affecting the atmospheric composition of radiation laboratories). *Atomnaya Energiya*, vol. 16, No. 3, pp. 282–283.

90. Dmitriev, M.T. 1965. Vliyanie ioniziruyushchie radiatsii na sostav vozdukha proizvodstvennykh pomeshchenii (The effect of ionizing radiation on the air composition on the production premises). *Gigiena i Sanitariya*, vol. 4, pp. 39–41.

91. Dmitriev, M.T. 1970. Prakticheskie aspekty radiatsionnoi khimii vozdukha (Practical aspects of radiation chemistry in the air). *Izotopy v SSSR*, No. 17, pp. 11–16.

92. *Dozimetriya Intensivnykh Potokov Ioniziruyushchikh Izluchenii* (Dosimetry of intensive currents of ionizing radiation). 1969. "Fan" Publishers, Tashkent.

93. Doroshenko, L.V. 1930. Vliyanie rentgenizatsii na dlinu vegetatsivnogo perioda u rastenii (The effects of x-rays on the length of the vegetative period of plants). In the book: *Trudy Instituta Prikladnoi Botaniki, Genetiki i Selektsii*. Lengiz, Leningrad, vol. 23, pp. 511–520.

94. Drobkov, A.A. 1962. Radioaktivnye elementy i urozhai (Radioactive elements and yield). Zemledelie, No. 8, pp. 24–35.

95. Drobkov, A.A. 1958. *Mikroelementy i Estestvennye Radioaktivnye Elementy v Zhizni Rastenii i Zhivotnykh* (Microelements and natural radioactive elements in the lives of plants and animals). Izd-vo AN SSSR, Moscow.

96. Zhanova, T.P. 1970. Predposevnoe obluchenie semyan morkovi v proizvodstvennykh usloviyakh (Presowing irradiation of carrot seeds under commercial cultivation conditions). In the book: *Materialy Pervoi Nauchno-prakticheskoi Konferentsii Po Primeneniyu Izotopov i Ioniziruyushchikh Izluchenii v Sel'skom Khozyaistve*. Kishinev, pp. 23–24.

97. Zhezhel', N.G. 1957. Vliyanie radioaktivnykh veshchestv na urozhainost' i nekotorye biokhimicheskie protsessy v pochve i rasteniyakh pri vyrashchivanii kukuruzy (The effect of radioactive substances on yield and some biochemical processes in the soil and plants during corn cultivation). In the book: *Trudy Mezhuzovskogo Soveshchaniya Po Mikroelementam*. Lenizdat, Leningrad, pp. 215–221.
98. Zhezhel', N.G. 1957. O napravlenii nauchno-issledovatel'skikh rabot po primeneniyu radioaktivnykh veshchestv v biologii i sel'skom Khozyaistve (About the direction of research on the use of radioactive substances in biology and agriculture). In the book: *Trudy Mezhuzovskogo Soveshchaniya Po Mikroelementam*. Lenizdat, Leningrad, pp. 240–244.
99. Zhezhel', N.G. 1955. Vliyanie estestvennykh radioaktivnykh veshchestv na urozhainost' sel'skokhozyaistvennykh rastenii (The effect of natural radioactive substances on the yield of agricultural plants). In the book: *Sessiya AN SSSR Po Mirnomu Ispol'zovaniyu Atomnoi Energii*. Izd-vo AN SSSR, Moscow, pp. 48–51.
100. Zhezhel', N.G. 1956. Osnovnye itogi issledovanii po primeneniyu slantsevoi muki v kachestve Organo-mineral'nogo radioaktivnogo udobreniya (Main results of research on the use of schistose meal as organo-mineral radioactive fertilizer). *Zap. Leningradsk. S.kh. In-ta*, No. 11, pp. 103–106.
101. Zhezhel', N.G. 1958. O mekhanizme biologicheskogo deistviya malykh doz ioniziruyushchikh izluchenii v rasteniyakh (Biological actions of low doses of ionizing radiation in the plants). *Vestn. S.kh. Nauki*, No. 8, pp. 32–35.
102. Zhezhel', N.G. 1963. Effektivnost' predposevenoi obrabotki semyan yachmenya, pshenitsy i kukuruzy gamma-luchami ^{60}Co (The effectiveness of presowing treatment of the barley, wheat and corn seeds by ^{60}Co gamma-rays). In the book: *Predposevenoe Obluchenie Semyan Sel'skokhozyaistvennykh Kul'tur*. Izd-vo AN SSSR, Moscow, pp. 174–178.
103. Zhezhel', N.G. 1972. O faktorakh, modifitsiruyushchikh effektivnost' predposevnogo gamma-oblucheniya semyan (About the factors modifying the effectiveness of presowing gamma-irradiation of seeds). In the book: *Tezisy Dokladov Vsesoyuznoi Konferentsii Po Ispol'zovaniyu Radiatsionnoi Tekhniki v Sel'skom Khozyaistve*, Kishinev, vol. 1, p. 130.
104. Zaitsev, B.A. and A.I. Grivkova. 1961. *Radioaktivnyi Tsezii-137* (Radioactive ^{137}Cs). Gosatomizdat, Moscow.
105. *Zashchita Transportnykh Ustanovok s Yadernym Dvigatelem* (Protection of Mobile Units with a Nuclear Engine). 1961. Translation from English. Izd-vo inostr. lit. Moscow.

106. Zashchita Yadernykh Reaktorov (Protection of nuclear reactors). 1958. Materials of the Atomic Energy Commission, USA. Translation from English. Izd-vo inostr. lit., Moscow.
107. Zeinalov, I.I., A.O. Aliev and R.R. Riza-Zade. 1972. Vliyanie predposevnogo gamma-oblucheniya na urozhai i fitoftoroz Kartofelya (The effect of presowing gamma-irradiation on the yield and phytophthora blight of potato). *Radiobiologiya*, vol. 12, No. 2, pp. 311–315.
108. Ibragimov, Sh.I. 1963. Deistvie predposevnogo oblucheniya semyan gamma-luchami ^{60}Co na rost i razvitie Khlopchatnika (The effect of presowing irradiation of seeds by ^{60}Co gamma-rays on the growth and development of cotton). In the book: *Predposevnoe Obluchenie Semyan Sel'skokhozyaistvennykh Kul'tur*. Izd-vo AN SSSR, Moscow, pp. 204–210.
109. Ibragimov, Sh.I. and R.I. Koval'chuk. 1963. Vliyanie oblucheniya rastenii Khlopchatnika na razlichnykh stadiyakh ikh razvitiya (The effects of the irradiation of cotton plants at different stages of development). In the book: *Predposevnoe Obluchenie Semyan Sel'skokhozyaistvennykh Kul'tur*. Izd-vo AN SSSR, Moscow, pp. 170–173.
110. Ibragimov, Sh.I. 1971. Rezul'taty issledovanii i perspektivy dal'neishikh rabot po predposevnomu oblucheniyu semyan Khlopchatnika (Results of research and prospects of further work on the presowing irradiation of cotton seeds). In the book: *Deistvie Radiatsii Na Rasteniya*. "Fan" Publishers, Tashkent, pp. 6–8.
111. Ibragimov, Sh.I. and R.I. Koval'chuk. 1972. Povyshenie urozhaya Khlopchatnika putem predposevnogo oblucheniya semyan gamma-luchami ^{60}Co (Increase of cotton yield through presowing irradiation of seeds by ^{60}Co gamma-rays). In the book: *Tezisy Dokladov Vsesoyuznoi Konferentsii Po Ispol'zovaniyu Radiatsionnoi Tekhniki v Sel'skom Khozyaistve*. Kishinev, vol. 1, pp. 132–133.
112. Ivanov, V.I. 1970. *Kurs Dozimetry Uchebnik Dlya Vuzov* (Dosimetry. A textbook for degree courses). Second revised and enlarged edition. Atomizdat, Moscow.
113. *Instruktivnye Materialy Po Metodam Opredeleniya v Vozdukhe Robochikh Pomeshchenii Nekotorykh Khimicheskikh Veshchestv* (Instructional materials on methods to determine some chemical substances in the air on the working premises). 1964. "Meditsina" Publishers, Moscow.
114. Kabakchi, A.M., Ya.I. Lavrentovich and V.V. Pen'kovskii. 1963. *Khimicheskaya Dozimetriya Ioniziruyushchikh Izluchenii* (Chemical dosimetry of ionizing radiation). Izd-vo AN UkrSSR, Kiev.
115. Kabulov, D.T. 1960. Deistvie radioaktivnykh izotopov na rost, razvitie i urozhainost' Khlopchatnika (The effect of radioactive isotopes on the growth, development and yield of cotton). *Khlopkovodstvo*, No. 8, pp. 45–48.

116. Kazakov, Yu.I., L.I. Shemenkov and G.P. Chugunov. 1972. Primenenie nekotorykh statsionarnykh gamma-ustanovok v sel'skom khozyaistve (The use of some stationary gamma-units in agriculture). In the book: *Tezisy Dokladov Vsesoyuznoi Konferentsii Po Ispol'zovaniyu Radiatsionnoi Tekhniki v Sel'skom Khozyaistve.* Kishinev, vol. 3, pp. 25–26.
117. Kaushanskii, D.A. 1971. Peredvizhnaya gamma-ustanovka "Stimulyator" (The mobile gamma-unit "Stimulyator"). *Atomnaya Energiya,* vol. 30, No. 5, pp. 479–481.
118. Kaushankii, D.A. 1970. Sovremennoe sostoyanie i perspektivy razvitiya izotopnoi radiatsionnoi tekhniki i metodov dozimetrii dlya predposevnogo oblucheniya semyan sel'skokhozyaistvennykh rastenii (The contemporary status and developmental prospects of isotope radiation technology and dosimetric methods for the presowing irradiation of agricultural plant seeds). In the book: *Materialy Pervoi Nauchno-prakticheskoi Konferentsii Po Primeneniyu Izotopov i Ioniziruyushchikh Izluchenii v Sel'skom Khozyaistve.* Kishinev, pp. 19–20.
119. Kaushanskii, D.A. 1972. Razrabotka novoi radiatsionnoi tekhniki dlya sel'skokhozyaistvennogo proizvodstva (Development of a new radiation technology for agricultural production). In the book: *Tezisy Dokladov Vsesoyuznoi Konferentsii Po Ispol'zovaniyu Radiatsionnoi Tekhniki v Sel'skom Khozyaistve.* Kishiniv, vol. 1, pp. 8–12.
120. Kaushanskii, D.A. 1970. *Laboratornaya Gamma-Ustanovka Dlya Mikrobiologicheskikh i Biokhimicheskikh Issledovanii LMB-γ-1M* (The laboratory gamma-unit for microbiological and biochemical research). Atomizdat, Moscow.
121. Kaushanskii, D.A. 1970. Sistemy zaryadki i smeny istochnikov izlucheniya obluchatelei gamma-ustanovok (Systems for loading and replacing the radiation sources of gamma-unit irradiators). *Izotopy v SSSR,* No. 17, pp. 32–35.
122. Kaushanskii, D.A. and A.V. Antonovich, 1972. Peredvizhnaya proizvodsvennaya gamma-ustanovka "Kolos-M" (Mobile commercial gamma-unit "Kolos-M"). In the book: *Tezisy Dokladov Vsesoyuznoi Konferentsii Po Ispol'zovaniyu Radiatsionnoi Tekhniki v Sel'skom Khozyaistve.* Kishinev, vol. 3, pp. 17–18.
123. Kaushanskii, D.A. and Ya.A. Gurevich. 1969. Ustroistvo dlya oblucheniya biologicheskikh ob"ektov (An equipment for the irradiation of biological objects). Authorship (patent) certificate No. 324773 with priority dated July 29, 1969.
124. Kaushanskii, D.A. and B.G. Zhukov. 1970. Peredvizhnaya proizvodstvennaya gamma-ustanovka "Kolos" (Mobile commercial gamma-unit "Kolos"). *Atomnaya Energiya,* vol. 28, No. 4, pp. 366–367.
125. Kaushanskii, D.A. and B.G. Zhukov. 1970. Proizvodsvennaya pered-

vizhnaya gamma-ustanovka "Kolos" (Commercial mobile gamma-unit "Kolos"). In the book: *Materialy Pervoi Nauchno-prakticheskoi Konferentsii Po Primeneniyu Izotopov i Ioniziruyushchikh Izluchenii v Sel'skom Khozyaistve*. Kishinev, p. 45.

126. Kaushanskii, D.A. and Li Don Khva. 1969. Dozimetriya pri nepreryvnykh proizvodstvennykh protsessakh radiatsionnoi obrabotki sel'skokhozyaistvennykh kul'tur (Dosimetry in the continuous production processes of radiation treatment of agricultural crops). *Atomnaya Energiya*, vol. 27, No. 3, pp. 253–254.

127. Kaushanskii, D.A. and Li Don Khva. 1972. O vozmozhnosti ispol'zovaniya razlichnykh vidov termolyuminestsentnykh detektorov dlya dozimetrii proizvodstvennykh protsessov predposevnogo oblucheniya (About the possibility of using different types of thermoluminescent detectors to monitor the production processes of presowing irradiation). In the book: *Tezisy Dokladov Vsesoyuznoi Konferentsii Po Ispol'zovaniyu Radiatsionnoi Tekhniki v Sel'skom Khozyaistve*. Kishinev, vol. 3, pp. 36–37.

128. Kaushanskii, D.A., O.F. Partolin and E.D. Chistov. 1972. Ispol'zovanie spektral'nykh i dozovykh kharakteristik gamma-izlucheniya, proshedshego cherez mnogostupenchatyi tekhnologicheskii kanal v zashchite, prednaznachennyi dlya nepreryvnogo protsessa oblucheniya sel'skokhozyaistvennoi produktsii (The use of spectral and dose characteristics of gamma radiation that had passed through a multistep, technological canal in the protection designed for the continuous process of irradiating agricultural produce). In the book: *Tezisy Dokladov Vsesoyuznoi Konferentsii Po Ispol'zovaniyu Radiatsionnoi Tekhniki v Sel'skom Khozyaistve*. Kishinev, vol. 3, pp. 21–23.

129. Kaushanskii, D.A., B.M. Terent'ev and S.K. Dubnova. 1972. Issledovanie i raschet parametrov obluchatelei slozhnykh konfiguratsii dlya izotopnykh gamma-ustanovok v sel'skom khozyaistve (Study and calculation of the parameters for irradiators of complex configuration for the isotope gamma-units in agriculture). In the book: *Tezisy Dokladov Vsesoyuznoi Konferentsii Po Ispol'zovaniyu Radiatsionnoi Tekhniki v Sel'skom Khozyaistve*. Kishinev, vol. 3, pp. 20–21.

130. Kedrov-Zikhman, O.K., A.F. Agafonova and A.N. Kozhevnikova. 1957. Vliyanie gamma-izlucheniya ^{60}Co na sel'skokhozyaistvennye rasteniya (The effect of ^{60}Co gamma-radiation on agricultural plants). In the book: *Tezisy Dokladov Na Vsesoyuznoi Nauchnotekhnicheskoi Konferentsii Po Primeneniyu Radioaktivnykh i Stabil'nykh Izotopov i Izluchenii v Narodnom Khozyaistve i Nauke*. Izd-vo AN SSSR, Moscow, p. 56.

131. Kedrov-Zikhman, O.K. and N.I. Borisova. 1963. Deistvie predposevnogo oblucheniya semyan gamma-luchami ^{60}Co na sel'skokhozyaist-

vennye rasteniya (The effects of presowing irradiation of agricultural plant seeds by ^{60}Co gamma-rays). In the book: *Predposevnoe Obluchenie Semyan Sel'skokhozyaistvennykh Kul'tur*. Izd-vo AN SSSR, Moscow, pp. 119–125.

132. Kietse, V.T. 1960. Vliyanie gamma-radiatsii i izmeneniya pitaniya zarodisha na rost i razvitie kukuruzhy (The effects of gamma-radiation and embryonic nutritional changes on the growth and development of corn). Ph.D. thesis for a Candidate of Biological Sciences degree, Institute of Biology of the Academy of Sciences of LatvSSR, Riga.
133. Kietse, V.T. 1961. Predposevnoe obluchenie semyan kukuruzy i ego prakticheskaya tsennost' (Presowing irradiation of corn seeds and its practical importance). In the book: *Nauka Sel'skomu Khozyaistvu*. Izd-vo AN LatvSSR, Riga, pp. 63–68.
134. Kietse, V.T. 1960. Vliyanie izluchenii ^{60}Co na meristematicheskie kletki kukuruzy (The effects of ^{60}Co radiation on meristematic corn cells). Izv. AN LatvSSR, No. 5, pp. 149–153.
135. Kietse, V.T. 1959. Vliyanie oblucheniya radioaktivnym kobal'tom na prorastanie semyan i rost kukuruzy (Effects of irradiation by radioactive cobalt on seed germination and the growth of corn plants). *Izv. AN LatvSSR*, No. 5, pp. 131–135.
136. Klechkovskii, V.M. 1957. *Izotopy i Izlucheniya v Agronomii* (Vsesoyuznaya nauchno-tekhnicheskaya konferentsiya po primeneniyu radioaktivnykh i stabil'nykh izotopov i izluchenii v narodnom khozyaistve i nauke) (Isotopes and radiation in agronomy. All-union science and technical conference on the use of radioactive and stable isotopes and radiation on the national economy and science). Izd-vo AN SSSR, Moscow.
137. Kokroft, D. 1958. Perspektivy ispol'zovaniya atomnoi energii (The prospects of using atomic energy). *Atomnaya Energiya*, vol. 4, pp. 417–418.
138. Kopylov, V.A. and A.M. Kuzin. 1971. Ob izmenenii fiziko-khimicheskikh svoistv polifenoloksidazy i ego izologov v gamma-obluchennykh rastitel'nykh tkanyakh (About the change of physical and chemical properties of polyphenoloxidase and its isologs in gamma-irradiated plant tissues). *Dokl. AN SSSR, Biol.*, vol. 196, No. 4, pp. 965–968.
139. Kostin, V.I. 1970. Vliyanie predposevnogo oblucheniya na rost, razvitie i biokhimicheskii sostav korneplodov sakharnoi svekly (The effect of presowing irradiation on the growth, development and biochemical composition of sugarbeet tubers). In the book: *Materialy Pervoi Nauchno-prakticheskoi Konferentsii Po Primeneniyu Izotopov i Ioniziruyushchikh Izluchenii v Sel'skom Khozyaistve*. Kishinev, pp. 34–35.
140. Kostin, V.I. 1971. Deistvie predposevenogo oblucheniya semyan na urozhai i tekhnologicheskii sostav sakharnoi svekly v polevykh i pro-

izvodstvennykh usloviyakh (The effect of presowing irradiation of seeds on the yield and technological composition of sugarbeet under field and large-scale commercial cultivation conditions). In the book: *Deistvie Radiatsii Na Rasteniya.* "Fan" Publishers, Tashkent, pp. 12–13.

141. Kostin, V.I. 1972. Osobennosti rosta sakharnoi svekly pri deistvii gamma-luchei (The peculiarities of sugarbeet growth due to the effects of gamma rays). In the book: *Tezisy Dokladov Vsesoyuznoi Konferentsii Po Ispol'zovaniyu Radiatsionnoi Tekhniki v Sel'skom Khozyaistve,* Kishinev, vol. 1, pp. 67–68.

142. Kryukova, L.M., A.M. Kuzin and K.S. Listvin. 1963. Predposevnaya obrabotka semyan l'na-dolguntsa gamma-luchami s tsel'yu stimulyatsii rosta rastenii i uluchcheniya kachestva volokna (Presowing treatment of flax fiber seeds by gamma rays to stimulate plant growth and improve the fiber quality). In the book: *Predposevnoe Obluchenie Semyan Sel'skokhozyaistvennykh Kul'tur.* Izd-vo AN SSSR, Moscow, pp. 89–93.

143. Kuzin, A.M. 1958. O nachal'nykh mekhanizmakh biologicheskogo deistviya ioniziruyushchikh izluchenii (About the initial biological effects of ionizing radiation). In the book: *Vsesoyuznaya Konferentsiya Po Primeneniyu Izotopov i Yadernykh Izluchenii Radiobiologiya.* Izd-vo AN SSSR, Moscow, pp. 3–13.

144. Kuzin, A.M. 1963. K teorii predposevnogo gamma-oblucheniya semyan (The theory of presowing gamma irradiation of seeds). In the book: *Seminar Po Primeneniyu Istochnikov Yadernykh Izluchenii Dlya Povysheniya Urozhainosti Sel'skokhozyaistvennykh Kul'tur i Polucheniya Novykh Khozyaistvennotsennykh Form Rastenii.* VDNKh, Moscow, p. 3.

145. Kuzin, A.M. 1972. Molekulyarnye mekhanizmy stimuliruyushchego deistviya ioniziruyushchikh izluchenii na semena rastenii (Molecular mechanisms of the stimulation effect of ionizing radiation on plant seeds). *Radiobiologiya,* vol. 12. No. 5, pp. 635–643.

146. Kuzin, A.M., N.M. Berezina and O.N. Shlykova. 1960. K voprosu o vliyanii moshchnosti dozy na radiobiologicheskii effekt u rastenii (The problem of the effects of dosage strength on the biological effect in plants). *Radiobiologiya,* vol. 5, No. 5, pp. 566–569.

147. Kuzin, A.M. and Yu.N. Runova. 1967. O postradiatsionnom vosstanovlenii semyan yachmenya (Post-radiation restoration of barley seeds). *Radiobiologiya,* vol. 7, No. 6, pp. 909–912.

148. Lur'e, L.S. and others. 1965. Obluchatel'nye ustanovki vo VNII elektrifikatsii sel'skogo khozyaistva (Radiation units in the All-Union Research Institute of Electrification of Agriculture). *Atomnaya Energiya,* vol. 19, No. 2, pp. 174–175.

149. Luchnik, N.V. 1960. Problema radiostimulyatsii rastenii i tsitologiche-

skii analiz yavleniya radiostimulyatsii (The problem of radiostimulation of plants and a cytological analysis of the phenomenon of radiostimulation). *Trudy In-ta Biol. Ural'sk. Filiala AN SSSR*, vol. 12, pp. 139–143.

150. Luchnik, N.V. 1952. O vliyanii fraktsionirovaniya i moshchnosti dozy na tsitologicheskii effekt oblucheniya (The effects of fractioning and dosage strength on the cytological effect of irradiation). *Biofizika*, vol. 1, No. 7, pp. 652–656.

151. Lysikov, V.N. and others. 1970. Itogi proizvodstvennoi aprobatsii metoda predposevnogo oblucheniya semyan v usloviyakh Moldavskoi SSR (Results of a large-scale, commercial trial of the method of presowing irradiation of seeds under conditions in the Moldavian SSR). In the book: *Materialy Pervoi Nauchno-prakticheskoi Konferentsii Po Primeneniyu Izotopov i Ioniziruyushchikh Izluchenii v Sel'skom Khozyaistve*. Kishinev, pp. 17–18.

152. Lysikov, V.N. and others. 1971. Trekhletnii rezul'taty proizvodstvennogo ispytaniya metoda predposevnogo obluchaniya semyan na peredvizhnoi gamma-ustanovke "Kolos" (Three-year results of a large-scale commercial trial of the method of presowing irradiation of seeds on the mobile gamma-unit "Kolos"). In the book: *Deistvie Radiatsii Na Rasteniya*. "Fan" Publishers, Tashkent, pp. 10–11.

153. Lemb, E. and others. 1959. Poluzavodskaya ustanovka dlya pererabotki produktov deleniya i drugie meropriyatiya programmy rabot po radioaktivnym izotopam v Ok-Ridzhskoi natsional'noi laboratorii (Semi-commercial unit for processing the products of division and other measures of the research program on radioactive isotopes at the Oak Ridge national laboratory). In the book: *Trudy Vtoroi Mezhdunarodnoi Konferentsii Po Mirnomu Ispol'zovaniyu Atomnoi Energii. Geneva, 1958. Poluchenie i Primenenie Izotopov*. Selected works of foreign scientists. Atomizdat, Moscow, vol. 10, p. 115.

154. Malishchuk, I.Ya. 1972. Voprosy dozimetrii i effektivnost' biofizicheskogo vliyaniya rentgenovskoi radiatsii na khod vegetatsii i produktivnost' kukuruzy (Monitoring problems and the biophysical effects of x-radiation on the vegetative growth and productivity of corn). In the book: *Tezisy Dokladov Vsesoyuznoi Konferentsii Po Ispol'zovaniyu Radiatsionnoi Tekhniki v Sel'skom Khozyaistve*. Kishinev, vol. 1, p. 29.

155. Mal'kov, I.A., V.V. Nikol'skii and D.A. Kaushanskii. 1969. Radiatsionnaya bezopasnost' na moshchnykh gamma-ustanovkakh s nepodvizhnymi obluchatelyami (Radiation safety on powerful gamma-units with stationary irradiators). In the book: *Nauchnye Raboty Institutov Okhrany Truda VTsSPS*. Profizdat, Moscow, vol. 57, pp. 47–50.

156. Malyarenko, S.G. 1963. Vliyanie ioniziruyushchikh izluchenii na prorastanie semyan, rost i razvitie seyantsev nekotorykh sortov lavandy

(The effect of ionizing radiation on seed germination and the growth and development of seedlings of some lavender varieties). In the book: *Predposevnoe Obluchenie Semyan Sel'skokhozyaistvennykh Kul'tur*. Izd-vo AN SSSR, Moscow, pp. 161–163.

157. Martem'yanov, Yu.A. 1971. Metody i rezultaty nachal'nykh eksperimentov po predposevnomu oblucheniyu semyan nekotorykh sel'skokhozyaistvennykh kul'tur v usloviyakh Pavlodarskogo Priirtysh'ya (Methods and results of initial experiments on the presowing irradiation of some agricultural crop seeds under conditions in Pavlodar province). In the book: *Deistvie Radiatsii Na Rasteniya*. "Fan" Publishers, Tashkent, pp. 13–14.

158. Martem'yanov, Yu.A. and A.V. Kalashnikov. 1972. Primenenie peredvizhnykh gamma-ustanovok "Kolos" dlya povysheniya urozhainosti sel'skokhozyaistvennykh kul'tur v usloviyakh Pavlodarskoi oblasti (The use of the mobile gamma-units "Kolos" to increase the yield of agricultural crops under the conditions in the Pavlodar province). In the book: *Tezisy Dokladov Vsesoyuznoi Konferentsii Po Ispol'zovaniyu Radiatsionnoi Tekhniki v Sel'skom Khozyaistve*. Kishinev, vol. 1, p. 37.

159. Materialy Pervoi Nauchno-prakticheskoi Konferentsii po Radiatsionnoi Bezopasnosti, 23–29 Noyabrya, 1966 (Proceedings of the First Scientific and Practical Conference on Radiation Safety, November 23–29, 1966), Moscow, 1968.

160. Materialy Vtoroi Nauchno-prakticheskoi Konferentsii Po Radiatsionnoi Bezopasnosti, 14–18 Sentyabrya, 1970 (Proceedings of the Second Scientific and Practical Conference on Radiation Safety, September 14–18, 1970), Moscow, 1970.

161. Mashkovich, V.P. and V.A. Klimanov. 1966. Raspredelenie intensivnosti gamma-izlucheniya v polom pryamom tsilindiricheskom kanale (Distribution intensity of gamma-radiation in a straight, hollow, cylindrical canal). *Atomnaya Energiya*, No. 20, pp. 127–128.

162. Medvedev, L.I. 1967. *Gigiena Truda Pri Rabote s Yadokhimikatami* (Hygiene while working with poisonous chemicals). "Meditsina" Publishers, Moscow.

163. Metodicheskie Ukazaniya Po Predposevnomu Gamma-oblucheniyu Semyan Sel'skokhozyaistvennykh Rastenii (Methodical instructions on presowing gamma-irradiation of agricultural plant seeds). Authors: N.M. Berezina, R.R. Riza-Zade, D.A. Kaushanskii and others. Atomizdat, Moscow, 1970.

164. Metodicheskie Ukazaniya Po Predposevnomu Gamma-oblucheniyu Semyan Sel'skokhozyaistvennykh Rastenii (Methodical instructions on presowing gamma-irradiation of agricultural plant seeds). Authors: N.M. Berezina, N.F. Batygin, D.A. Kaushanskii and others. Kishinev Agricultural Institute, Kishinev, 1972.

165. Metodicheskie Ukazaniya Po Opredeleniyu Ekonomicheskoi Effektivnosti Metoda Predposevnogo Gamma-oblucheniya Semyan Sel'skokhozyaistvennykh Kul'tur i Vnedreniya Radiatsionnoi Tekhniki v Sel'skom Khozyaistve (Methodical instruction on the determination of the economic effectiveness of the method of presowing gamma-irradiation of agricultural crop seeds and the wider application of radiation technology in agriculture). Editor: G.Ya. Rud', Member-correspondent of the Academy of Sciences MSSR. Authors: D.A. Kaushanskii and A.P. Koval'. Kishinev, 1973.

166. Metodicheskie Ukazaniya No. 336 "Metodika Attestatsii Izotopnykh Gamma-ustanovok Po Moshchnosti ekspozitsionnoi Dozy" (Methodical instructions No. 336 "Method for the certification of the isotope gamma-units on the basis of the strength of exposure dose"). Authors: V.V. Generatova, M.N. Gurskii and D.A. Kaushanskii. "Standarty" Publishers, Moscow, 1973.

167. Metodicheskie Ukazaniya Po Opredeleniyu Effektivnosti Udobrenii v Sel'skom Khozyaistve (Methodical instructions for the determination of the effectiveness of fertilizers in agriculture). "Kolos" Publishers, Moscow, 1971.

168. Miller, A.T. 1959. Vozdeistvie beta-izluchatelei na rasteniya pri zamachivanii semyan v rastvorakh fosfora-32 (The effect of beta-radiations on plants after soaking seeds in phosphorus-32 solutions). *Vestn. S.Kh. Nauki*, No. 11, pp. 18–21.

169. Miller, A.T. 1970. Predposevnoe obluchenie semyan sakharnoi svekly i perspektivy vnedreniya metoda v usloviyakh Latvii (Presowing irradiation of sugarbeet seeds and prospects for the wider application of this method under Latvian conditions). In the book: *Materialy Pervoi Nauchno-prakticheskoi Konferentsii Po Primeneniyu Izotopov i Ioniziruyushchikh Izluchenii v Sel'skom Khozyaistve*. Kishinev, p. 33.

170. Miller, A.T. 1972. Effektivnosti' predposevnogo oblucheniya semyan sakharnoi svekly v usloviyakh Latviiskoi SSR (Effectiveness of the presowing irradiation of sugarbeet seeds under conditions in the Latvian SSR). In the book: *Tezisy Dokladov Vsesoyuznoi Konferentsii Po Ispol'zovaniyu Radiatsionnoi Tekhniki v Sel'skom Khozyaistve*. Kishinev, vol. 1, pp. 38–39.

171. Mur, R. and R. Berns. 1959. Vydelenie produktov deleniya iz radioaktivnykh sbrosnykh rastvorov (Separation of the products of division from radioactive waste solutions). In the book: *Trudy Vtoroi Mezhdunarodnoi Konferentsii Po Mirnomu Ispol'zovaniyu Atomnoi Energii*. Atomizdat, Moscow, vol. 5, pp. 376–378.

172. Mukhanova, V.L. 1971. Deistvie predposevnogo oblucheniya semyan na urozhai i kachestvo sena lyutserny (The effect of presowing irradiation of seeds on the yield and quality of lucerne hay). In the book:

Deistvie Radiatsii Na Rasteniya. "Fan" Publishers, Tashkent, pp. 8–9.

173. Nigmanov, A. 1971. Vliyanie predposevnogo oblucheniya na soderzhanie sinil'noi Kisloty v nekotorykh sel'skokhozyaistvennykh kul'turakh (The effect of presowing irradiation on the hydrocyanic acid content in some agricultural crops). In the book: *Deistvie Radiatsii Na Rasteniya.* "Fan" Publishers, Tashkent, p. 36.
174. Normy Radiatsionnoi Bezopasnosti (NRB-69) (Norms of Radiation Safety. NRB-69). Second edition. Atomizdat, Moscow, 1972.
175. Osnovnye Normy Bezopasnosti Pri Zashchite Ot Izluchenii (Basic Safety Norms of Protection against Radiation). MAGATE, Vienna, 1968.
176. Osnovnye Sanitarnye Pravila Raboti s Radioaktivnymi Veshchestvami i Drugimi Istochnikami Ioniziruyushchikh Izluchenii OSP-72 (Basic sanitary rules of working with radioactive substances and other sources of ionizing radiation OSP-72). Atomizdat, Moscow, 1973.
177. Osnovy Radiatsionno-khimicheskogo Apparatostroeniya (Fundamentals of designing radio-chemical apparatus). Authors: A.Kh. Breger, Yu.I. Vainshtein, N.P. Syrkus and others. Atomizdat, Moscow, 1967.
178. Palamarchuk, A.O. 1959. Vliyanie oblucheniya semyan morkovy gamma-luchami radiya na urozhai korneplodov (The effects of irradiating carrot seeds by radium gamma-rays on the yield of tubers). *Selektsiya i Semenovodstvo,* No. 4, pp. 69–74.
179. Panchenko, A.M. 1963. Nekotorye dozimetricheskie kharakteristiki malogabaritnogo schetchika SBM-10 (Some dosimetric characteristics of the small meter SBM-10). *Atomnaya Energiya,* No. 14, p. 408.
180. Parfenov, V.T. 1971. Izmenenie azotnogo obmena v rasteniyakh kartofelya i kachestva urozhaya pod deistviem uskorennykh elektronov (The change of nitrogen metabolism in potato plants and the quality of the harvest due to the effects of accelerated electrons). In the book: *Diestvie Radiatsii Na Rasteniya.* "Fan" Publishers, Tashkent, pp. 31–32.
181. Parfenov, V.T. and N.M. Berezina. 1968. Rol' fiziologicheskogo sostoyaniya klubnei kartofelya v effekte stimulyatsii pri gamma-obluchenii (The role of the physiological condition of potato tubers in the stimulation effect during gamma-irradiation). In the book: *Vsesoyuznaya Nauchnaya Konferentsiya Po Primeneniyu Izotopov i Izluchenii v Sel'skom Khozyaistve.* VASKhNIL, Moscow, p. 13.
182. Petros'yants, A.M. 1972. *Ot Nauchnogo Poiska k Atomnoi Promyshlennosti* (From scientific search to atomic industry). Second revised and enlarged edition. Atomizdat, Moscow.
183. Polyakova, N.I. and L.A. Sergeeva. 1972. Vliyanie razlichnykh doz predposevnogo gamma-oblucheniya na produktivnost' yarovogo yach-

menya (The effect of different doses of presowing gamma-irradiation on the productivity of spring barley). In the book: *Tezisy Dokladov Vsesoyuznoi Konferentsii Po Ispol'zovaniyu Radiatsionnoi Tekhniki v Sel'skom Khozyaistve*. Kishinev, vol. 1, p. 145.

184. Popandopulo, B.Kh. 1956. Metodika Zootekhnicheskogo Analiza (Method of zootechnical analysis). Sel'khozgiz, Moscow.

185. Popov, V.I. 1962. Preobrazovanie ot tochki drugikh prosteishikh form istochnikov k tsilindru (Transformation from a point on other simple forms of sources to a cylinder). In the book: *Pribory i Metody Analiza Izluchenii*. Gosatomizdat, Moscow, vol. 3, p. 8.

186. Poryadkova, N.A. 1956. Metodika i rezul'taty nekotorykh spadov v radio-stimulyatsii rastenii (The method and results of some decreases in plant radio-stimulation). *Biofizika*, vol. 1, No. 7, pp. 597–603.

187. Poryadkova, N.A., N.M. Makarov and N.V. Timofeev-Resovskii. 1957. Opyty po radioaktivnoi stimulyatsii rastenii (Experiments on the radiostimulation of plants). In the book: *Trudy Vsesoyuznoi Konferentsii Po Meditsine i Radiologii*. Medgiz, Moscow, pp. 43–47.

188. Poryadkova, N.A., N.M. Makarov and N.V. Kulikov. 1960. Opyty po radiostimulyatsii kul'turnykh rastenii (Experiments on the radiostimulation of cultivated plants). *Trudy In-ta Biol. Ural'sk. Filiala AN SSSR*, vol. 3, pp. 68–71.

189. Poryadkova, N.A., N.V. Timofeev-Resovskii and N.V. Luchnik. 1960. Opyty Po obluchenіyu semyan gorokha i pshenitsy x- i gamma-luchami na raznykh stadiyakh zamachivaniya i prorastaniya (Experiments on irradiating seeds of pea and wheat by x- and gamma-rays at different stages of soaking and germination). *Trudy In-ta Biol. Ural'sk. Filiala AN SSSR*, vol. 3, p. 44.

190. Preobrazhenskaya, E.I. 1963. O korrelyatsii mezhdu stimuliruyushchim deistviem ioniziruyushchikh izluchenii i obshchei radiochuvstvitel'nost'-yu (Correlation between the stimulation effects of ionizing radiation and general radiosensitivity). In the book: *Predposevnoe Obluchenie Sel'skokhozyaistvennykh Kul'tur*. Izd-vo AN SSSR, Moscow, pp. 184–189.

191. Preobrazhenskaya, E.I. 1971. Radioustoichivost' Semyan Rastenii (Radioresistance of plant seeds). Atomizdat, Moscow.

192. Prikladnaya Dozimetriya (Applied dosimetry). 1962. Authors: K.K. Aglintsev, V.M. Kodyukov, A.F. Lyzlov and Yu.V. Sivintsev. Gosatomizdat, Moscow.

193. Prokof'ev, N.S. and D.A. Kaushanskii. 1968. Otsenka ekonomicheskoi effektivnosti ustanovski "Kolos" (Evaluation of economic effectiveness of the gamma-unit "Kolos"). Report at the 20th All-Union Science and Technical Conference "XX Let Proizvodstva i primeneniya Izotopov i Istochnikov Yadernykh Izluchenii v Narodnom Khozyaistve". Minsk.

194. Prokof'ev, N.S., D.A. Kaushanskii and B.G. Zhukov. 1970. Ekonomicheskaya effektivnost' ispol'z ovaniya proizvodstvennykh gamma-ustanovok "Kolos" (Economic effectiveness of the use of commercial gamma-units "Kolos"). *Atomnaya Energiya*, vol. 28, No. 5, pp. 453–454.

195. Prokhozhdenie Izluchenii Cherez Neodnorodnosti v Zashchite (Passage of radiation through Heterogeneous Protection). 1968. Authors: V.G. Zolotukhin, V.A. Klimanov, O.I. Leipunskii and others. Atomizdat, Moscow.

196. Raggenba, A. and others. 1959. Opytnaya ustanovka dlya vydeleniya tseziya-137 (Experimental unit for the isolation of Cesium-137). In the book: *Trudy Vtoroi Mezhdunarodnoi Konferentsii Po Mirnomu Ispol'zovaniyu Atomnoi Energii.* Atomizdat, Moscow, vol. 10, pp. 127–130.

197. Radiatsionnaya Dozimetriya (Radiation dosimetry). Translation from English. Izd-vo inostr. lit., Moscow, 1958.

198. Rapp, A.F. 1956. Metody obrabotki radioaktivnykh materialov v kolichestvakh mnogikh tysyach kyuri (Methods of treating radioactive materials in quantities of several thousand curie). In the book: *Dozimetriya Ioniziruyushchikh Izluchenii* (Reports of foreign scientists at the International Conference on the Peaceful Use of Atomic Energy. Geneva, 1955). Gostekhizdat, Moscow, pp. 543–545.

199. Rachinskii, V.V. 1970. Problemy ispol'zovaniya atomnoi tekhniki v sel'skom khozyaistve (The problems of using atomic technology in agriculture). In the book: *Materialy Pervoi Nauchno-prakticheskoi Konferentsii Po Primeneniyu Izotopov i Ioniziruyushchikh Izluchenii v Sel'skom Khozyaistve.* Kishinev, pp. 7–10.

200. Revin, A.F. 1971. Vliyanie predposevnogo oblucheniya semyan gamma-luchami ^{137}Cs na soderzhanie triptofana i serina v prorostkakh sel'skokhozyaistvennykh rastenii (The effects of the presowing irradiation of seeds by ^{137}Cs gamma-rays on the content of tryptophane and serine in agricultural plant seedlings). *Nauch. Dokl. Vyssh. Shkoly. Biol. Nauki.* No. 1, pp. 54–57.

201. Riza-Zade, R.R. 1970. Otsenka effektivnosti predposevnogo oblucheniya semyan v usloviyakh Azerbaidzhana i perspektivy vnedreniya (Evaluation of the effectiveness of the presowing irradiation of seeds under conditions in Azerbaijan and the prospects of its wider use). In the book: *Materialy Pervoi Nauchno-prakticheskoi Konferentsii Po Primeneniyu Izotopov i Ioniziruyushchikh Izluchenii v Sel'skom Khozyaistve.* Kishinev, pp. 25–27.

202. Rize-Zade, R.R. and Yu.V. Bozov. 1972. Kolichestvennye i kachestvennye slagaemye urozhaya klevera pri predposevnom obluchenii semyan (Quantitative and qualitative yield components of clover in the presowing irradiation of seeds). In the book: *Tezisy Dokladov Vseso-*

yuznoi Konferentsii Po Ispol'zovaniyu Radiatsionnoi Tekhniki v Sel'skom Khozyaistve. Kishinev, vol. 1, p. 140.

203. Rik, G.R. 1963. K voprosu o mekhanizme deistviya ioniziruyushchei radiatsii na rasteniya (The problem of the mechanisms of ionizing radiation action on plants). In the book: *Predposevnoe Obluchvnie Sel'skokhozyaistvennykh Kul'tur*. Izd-vo AN SSSR, Moscow, pp. 13–20.

204. Rodionov, T.A. 1972. Vliyanie predposevnogo oblucheniya semyan na urozhai i tekhnologicheskie kachestva korneplodov sakharnoi svekly (The effects of the presowing irradiation of seeds on the yield and technological qualities of sugarbeet tubers). In the book: *Tezisy Dokladov Vsesoyuznoi Konferentsii Po Ispol'zovaniyu Radiatsionnoi Tekhniki v Sel'skom Khozyaistve*. Kishinev, vol. 1, p. 141.

205. Roze, K.K. and G.E. Kavatse. 1963. Kratkovremennoe predposevnoe obluchenie semyan i dlitel'noe obluchenie rastenii gamma-luchami (Short-term presowing irradiation of seeds and long-term irradiation of plants by gamma-rays). In the book: *Predposevnoe Obluchenie Semyan Sel'skokhozyaistvennykh Kul'tur*. Izd-vo AN SSSR, Moscow, pp. 141–146.

206. Roze, K.K. and V.G. Kietse. 1963. Vliyanie oblucheniya radioaktivnym kobal'tom na rost i razvitie kukuruzi (The effect of irradiation by radioactive cobalt on the growth and development of corn). In the book: *Predposevnoe Obluchenie Semyan Sel'skokhozyaistvennykh Kul'tur*. Izd-vo AN SSSR, Moscow, pp. 131–140.

207. Savin, V.N. 1962. Izmenenie posledeistviya gamma-luchei v zavisimosti ot uslovii vyrashchivaniya rastenii (Changes in the after-effects of gamma-rays due to the cultivation conditions of plants). *Byul. Nauchn-tekhn. Inform. Po Agrofizike*, No. 10, pp. 11–15.

208. Savin, V.N. 1963. Vliyanie predposevnogo oblucheniya semyan malymi dozami luchei ^{60}Co na rost rastenii (The effect of presowing irradiation of seeds by low doses of ^{60}Co rays on the growth of the plants). In the book: *Predposevnoe Obluchevenie Semyan Sel'skokhozyaistvennykh Kul'tur*. Izd-vo AN SSSR, Moscow, pp. 190–193.

209. Sanitarnye Pravila Ustroistva i Ekspluatatsii Moshchnykh Izotopnykh Gamma-ustanovok, No. 482–64 (Sanitary rules of designing and using powerful gamma-units. No. 482–64). 1964. USSR Ministry of Health, Moscow.

210. Sanitarnye Pravila Ustroistva i Ekspluatatsii Radioaktivnykh Konturov Yadernykh Reaktorov, No. 654–66 (Sanitary rules of designing and using radioactive contours of nuclear reactors, No. 654–66). 1967. USSR Ministry of Health, Moscow.

211. Sanitarnye Pravila Ustroistva i Ekspluatatsii Moshchnykh Izotopnykh Gamma-ustanovok s Nepodvizhnym Obluchatelem, No. 774–68 (Sanitary rules of designing and using powerful isotope gamma units with

stationary irradiator, No. 774–68). 1969. Ministry of Health of the USSR, Moscow.

212. Satalkina, G.I. 1970. Sposoby polucheniya rannego urozhaya kaputsy v Tsentral'noi zone Krasnodarskogo Kraya (Methods of getting early cabbage harvests in central Krasnodar territory). Ph.D. thesis for the Candidate of Agricultural Sciences degree, Kuban Agricultural Institute, Krasnodar.

213. Serebrenikov, V.S. 1961. Vliyanie dlitel'nogo oblucheniya gamma-luchami na rasteniya Kartofelya (The effects of long-term gamma-irradiation on potato plants). In the book: *Trudy Nauchno-issledovatel'skogo Instituta Kartofel'nogo Khozyaistva*, Sel'khozgiz, Moscow, pp. 59–64.

214. Serebrenikov, V.S. 1962. Vliyanie razlichnykh doz gamma-luchei na rost, anatomo-morfologicheskoe stroenie rastenii, urozhai i kachestve kartofelya (The effects of different doses of gamma-rays on the growth, anatomical and morphological structure, yield and quality of potato plants). Ph.D. thesis for the Candidate of Agricultural Sciences degree. Research Institute of Potato Cultivation. Moscow.

215. Sinitsyn, V.I. 1967. Radioaktivnyi Kobal't—^{60}Co (Radioactive cobalt—^{60}Co). Second edition. Atomizdat, Moscow.

216. Sokurova, E.N. 1956. O nekotorykh zakonomernostyakh deistviya ioniziruyushchikh izluchenii na mikroorganizmy (Some rules of the effects of ionizing radiation on microorganisms). *Izv. AN SSSR, Ser. Biol.*, No. 6, pp. 35–53.

217. Sukach, K.I. 1962. Zavisimost' stimulyatsionnogo effekta ot moshchnosti rentgenovykh luchei pri obluchenii semyan kukuruzy (Correlation of stimulation effects with the strength of x-rays in the irradiation of corn seeds). In the book: *Materialy Nauchno-Otchetnoi Konferentsii Professorsko-prepodavatel'skogo Sostava Kishenevskogo Sel'skokhozyaistvennogo Instituta.* Izd-vo S.kh. lit., Kishinev, pp. 69–75.

218. Sukach, K.I. 1959. Vliyanie oblucheniya rentgenovskimi luchami semyan kukuruzy na ee produktivnost' (The effect of x-radiation on corn seeds and its productivity). In the book: *Trudy Ob''edinennoi Nauchnoi Sessii Moldavskogo Filiala AN SSSR*, Kishinev, vol. 2, pp. 135–139.

219. Sukach, K.I. and E.D. Morozova. 1959. Izmeneniya v khode morfogeneza kukuruzy v zavisimosti ot dozy ioniziruyushchego izlucheniya (Changes in the morphogenesis of corn depending on the ionizing radiation dose). In the book: *Trudy Ob''edinennoi Nauchnoi Sessii Moldavskogo Filiala AN SSSR*, Kishinev, vol. 2, pp. 159–161.

220. Sukach, K.I. and others. 1972. Issledovaniya problemy predposevnogo oblucheniya semyan v Moldavii (Research on the presowing irradiation of seeds in Moldavia). In the book: *Tezisy Dokladov Vsesoyuznoi Konferentsii Po Ispol'zovaniyu Radiatsionnoi Tekhniki v Sel'skom Khozyaistve*. Kishinev, vol. 1, pp. 5–7.

221. Suleimanova, I.G. 1972. Nekotorye storony proyavleniya effekta stimulyatsii pri predposevnom gamma-obluchenii semyan (Some aspects of the appearance of stimulation effects in the presowing gamma-irradiation of seeds). In the book: *Tezisy Dokladov Vsesoyuznoi Konferentsii Po Ispol'zovaniyu Radiatsionnoi Tekhniki v Sel'skom Khozyaistve*. Kishinev, vol. 1, pp. 143–144.

222. Sultanbaev, A.S. 1972. Predposevnoe obluchenie semyan sel'skokhozyaistvennykh kul'tur v Kirgizii (Presowing irradiation of agricultural crop seeds in Kirghizia). In the book: *Tezisy Dokladov Vsesoyuznoi Konferentsii Po Ispol'zovaniyu Radiatsionnoi Tekhniki v Sel'skom Khozyaistve*. Kishinev, vol. 1, p. 38.

223. Tezisy Chetvertogo Vsesoyuznogo Koordinatsionnogo Soveshchaniya Po Dozimetrii Intensivnykh Potokov Ioniziruyushchikh Izluchenii (Abstract of the reports at the all-union coordination meeting on the dosimetry of intensive ionizing radiation currents). Moscow, 1971.

224. Tekhnicheskie Usloviya Na Metody Opredeleniya Vrednykh Veshchestv v Vozdukhe (Technical conditions for methods to determine harmful substances in the air). "Meditsina" Publishers, Moscow, No. 3, 1964.

225. Timofeev-Resovskii, N.V. 1956. Biofizicheskaya interpretatsiya yavlenii radio-stimulyatsii rastenii (Biophysical interpretation of the phenomena of radiostimulation in plants). *Biofizika*, vol. 1, No. 7, pp. 616–621.

226. Timofeev-Resovskii, N.V. 1957. O deistvii slabykh doz ioniziruyushchikh izluchenii na rost i razvitie rastenii (The effects of low doses of ionizing radiation on the growth and development of plants). *Trudy In-ta Biol. Ural'sk. Filiala AN SSSR*, vol. 1, No. 9, pp. 57–60.

227. Timofeev-Resovskii, N.V. 1957. Tsitologicheskie i biofizicheskie osnovy radiostimulyatsii (Cytological and biophysical bases of radiostimulation). *Trudy In-ta Biol. Ural'sk. Filiala AN SSSR*, vol. 3, No. 13, pp. 5–11.

228. Timofeev-Resovskii, N.V. 1961. O podkhodakh k teoreticheskomu analizu mekhanizma radiostimulyatsii (Approaches for the theoretical analysis of the mechanisms of radio-stimulation). In the book: *Tezisy Dokladov Na Soveshchanii Po Predposevnomu Oblucheniyu Semyan Sel'skokhozyaistvennykh Kul'-tur*. Izd-vo AN SSSR, Moscow, p. 75.

229. Tipovaya metodika opredeleniya ekonomicheskoi effecktivnosti kapital'nykh vlozhenii novoi tekhniki v narodnom khozyaistve (A typical method to determine the economic effectiveness of capital investment and new technology in the national economy of the USSR). Izd-vo AN SSSR, Moscow, 1960.

230. Tipovaya metodika opredeleniya ekonomicheskoi effektivnosti (Typical method to determine economic effectiveness). Gosplan, Gosstroi, AN SSSR, 1969.

231. Trudy Vtorogo Koordinatsionnogo Soveshchaniya Po Dozimetrii Bol'-

shikh Doz (Works of the second coordination meeting on the dosimetry of high doses). "Fan" Publishers, Tashkent, 1966.

232. Trusova, V.P. and others. 1963. Ferrosul'fatnyi metod dozimetrii v metallicheskikh sosudakh (Ferrosulfate dosimetric method in metallic vessels). *Atomnaya Energiya*, vol. 5, pp. 6–7.

233. Tushnyakova, M.M. and M.A. Vasilevskii. 1966. Opyt po oblucheniyu semyan i Klubnei rastenii luchami Rentgena (Experiments on the x-radiation of plant seeds and tubers). *Izv. AN SSSR. ser. Biol.*, No. 1, pp. 157–159.

234. Fedorova, V.S. 1963. Vliyanie ioniziruyushchikh izluchenii na nakoplenie vitaminov i monosakhrov v list'yakh kukuruzy (The effect of ionizing radiation on the accumulation of vitamins and monosugars in corn leaves). In the book: *Predposevnoe Obluchenie Semyan Sel'skokhozyaistvennykh Kul'tur*. Izd-vo AN SSSR, Moscow, pp. 79–83.

235. Fedorova, V.S. and A.A. Sevast'yanova. 1957. Vliyanie ioniziruyushchikh izluchenii na biologicheskii aktivnye veshchestva v list'yakh kukuruzy (The effect of ionizing radiation on biologically active substances in corn leaves). In the book: *Tezisy Dokladov Na Vsesoyuznoi Nauchno-tekhnicheskoi Konferentsii Po Primeneniyu Radioaktivnykh i Stabil'nykh Izotopov i Izluchenii v Narodnom Khozyaistve i Nauke*. Izd-vo AN SSSR, Moscow, p. 54.

236. Fonshtein, L.M. and L.P. Chel'tsova. 1961. Vliyanie oblucheniya endosperma semyan pshenitsy na mitoticheskuyu aktivnost' kletok (The effects of irradiating wheat seed endosperm on the mitotic activity of cells). *Biologiya*, No. 4, pp. 619–623.

237. Frolov, G. 1936. Deistvie rentgenovskikh i ul'trafioletovykh luchei na rost rastenii (The effects of x-rays and ultraviolet rays on the growth of plants). *Trudy S.Kh. Akad. Im. K.A. Timiryazeva*, vol. 1, pp. 189–216.

238. Khmelev, B. and I. Malishchuk. 1960. Predposevnoe obluchenie semyan zhestkimi i myagkimi rentgenovskimi luchami (Presowing irradiation of seeds by severe and mild x-rays). *Nauch. Trudy Kamenets-Podol'skogo S.kh. In-ta*, No. 3, pp. 106–111.

239. Khudadatov, A.I. 1961. Nekotorye metodicheskie voprosy predposevnogo oblucheniya (Some methodical problems of presowing irradiation). *Izv. AN AzSSR. Ser. Biol.*, No. 12, pp. 33–35.

240. Khudadatov, A.I. 1963. Vliyanie radiatsii na rost i razvitie kukuruzy v usliviyakh Azerbaidzhana (The effects of radiation on the growth and development of corn under conditions in Azerbaijan). In the book: *Predposevnoe Obluchenie Semyan Sel'skokhozyaistvennykh Kul'tur*. Izd-vo AN SSSR, Moscow, pp. 199–203.

241. Khudadatov, A.I. 1963. Radiostimulyatsiya pri predposevnom obluchenii gamma-luchami vozdushnosukhikh semyan (Radiostimulation

in the presowing irradiation of air-dried seeds by gamma-rays). In the book: *Seminar Po Primeneniyu Istochnikov Yadernykh Izluchenii Dlya Povysheniya Urozhainosti Sel'skokhozyaistvennykh Kul'tur i Polucheniya Novykh Khozyaistvennotsennykh Form Rastenii*, VDNKh, Moscow, p. 8.

242. Tsuryopa, B.I. 1934. Vliyanie luchistoi energii X-luchei, ul'trafioletovykh luchei, radiovoln i tokob tesla na rost i razvitie nekotorykh kul'turnykh rastenii (The effect of the radiation energy of x-rays, ultraviolet rays, radio waves and Tesla points on the growth and development of some cultivated plants). In the book: *Sbornik Nauchno-issledovatel'skikh Rabot Azovo-Chernomorskogo Sel'skokhozyaistvennogo Instituta*. Rostov-on-Don, No. 3, pp. 24–32.

243. Chaikina, N.A. and S.N. Tselishchev. 1960. Tsitologo-embriologicheskie izmeneniya u gorokha vyrashchennogo iz semyan obluchennykh teplovymi neitronami (Cytological and embryological changes in pea grown from seeds irradiated by thermal neutrons). In the book: *Doklady Moskovskoi Sel'skokhozyaistvennoi Akademii Imeni K.A. Timiryazeva*, Moscow, No. 57, pp. 55–61.

244. Chemarin, N.G., A.I. Arinshtein and L.A. Kichanova. 1972. Biologicheskoe deistvie gamma-izluchenii na semena koriandra (Biological effects of gamma-radiation on coriander seeds). In the book: *Tezisy Dokladov Vsesoyuznoi Konferentsii Po Ispol'zovaniyu Radiatsionnoi Tekhniki v Sel'skom Khozyaistve*. Kishinev, vol. 1, pp. 102–103.

245. Chernov, N.P. 1972. Effektivnost' ispol'zovaniya predposevnogo gamma-oblucheniya semyan ogurtsov, tomatov dlya zakrytogo grunta (Effectiveness of the use of presowing gamma-irradiation of cucumber and tomato seeds for cultivation in glass-houses). In the book: *Tezisy Dokladov Vsesoyuznoi Konferentsii Po Ispol'zovaniyu Radiatsionnoi Tekhniki v Sel'skom Khozyaistve*. Kishinev, vol. 1, pp. 44–45.

246. Chekhov, N.V. 1934. Vliyanie x-luchei na rasteniya (The effect of x-rays on plants). *Trudy Tomsk. Un-ta*, vol. 85. pp. 62–135.

247. Chistov, E.D. and D.A. Kaushanskii. 1970. Osnovy radiatsionnoi bezopasnosti pri ispol'zovanii peredvizhnykh gamma-ustanovok v sel'skokhozyaistvennom proizvodstve (Fundamentals of radiation safety in the use of mobile gamma-units in agricultural production). In the book: *Materialy Pervoi Nauchno-prakticheskoi Konferentsii Po Primeneniyu Izotopov i Ioniziruyushchikh Izluchenii v Sel'skom Khozyaistve*. Kishinev, p. 46.

248. Chistov, E.D., D.A. Kaushanskii, I.A. Mal'kov and V.V. Nikol'skii. 1968. Issledovanie radiatsionnoi bezopasnosti gamma-ustanovok s nepodvizhnymi obluchatelyami v radiobiologii i sel'skom khozyaistve (Study of radiation safety of gamma-units with stationary irradiators in radiobiology and agriculture). Report at the All-Union Science-

Technological Conference "XX Let Proizvodstva i Primeneniya Izotopov i Istochnikov. Yadernykh Izluchenii v Narodnom Khozyaistve". Minsk.

249. Chistov, E.D., D.A. Kaushanskii, O.F. Partolin and N.D. Styrov. 1970. Spektrometricheskie kharakteristiki gamma-izlucheniya v mnogosektsionnykh tekhnologicheskikh kanalakh proizvodstvennykh izotopnykh ustanovok (Spectrometric features of gamma-radiation in the multisectional technological canals of commercial isotope units). In the book: *Materialy Vtoroi Nauchno-prakticheskoi Konferentsii Po Radiatsionnoi Bezopasnosti (14–18 September, 1970)*. Moscow, p. 30.

250. Chistov, E.D., O.F. Partolin and D.A. Kaushanskii. 1971. Issledovanie prokhozhdeniya gamma-izlucheniya cherez mnogosektsionnyi tekhnologicheskii kanal v zashchite (Soobshchenie 1) (Study on the passage of gamma-radiation through the multisectional technological canal for protection. Report 1). In the book: *Sbornik Nauchnykh Rabot Institutov Okhrany Truda VTsSPS*. Profizdat, Moscow, No. 70, pp. 36–40.

251. Chistov, E.D., O.F. Partolin, A.M. Yunin and D.A. Kaushanskii. 1972. Issledovanie prokhozhdeniya gamma-izlucheniya cherez mnogosektsionnyi tekhnologicheskii kanal v zashchite (Soobshchenie 2) (Study on the passage of gamma-radiation through the multisectional technological canal for protection. Report 2). In the book: *Sbornik Nauchnykh Rabot Institutov Okhrany Truda VTsSPS*. Profizdat, Moscow, No. 75, pp. 52–54.

252. Chistov, E.D. and others. 1968. Ob organizatsii i rabote sluzhby radiatsionnoi bezopasnosti na moshchnykh radiatsionno-khimicheskikh ustanovkakh (Organization and work of the radiation safety service on powerful radiation-chemical units). In the book: *Sbornik Nauchnykh Rabot Institutov Okhrany Truda VTsSPS*. Profizdat, Moscow, No. 53, pp. 50–52.

253. Chistova, K.N. 1970. Vliyanie gamma-oblucheniya na rost, razvitie i urozhai razlichnykh sortov yarovoi pshenitsy (Ul'yanovskaya Oblast') (The effect of gamma irradiation on the growth, development and yield of different varieties of spring wheat in the Ul'yanovsk province). In the book: *Materialy Pervoi Nauchno-prakticheskoi Konferentsii Po Primeneniyu Izotopov i Ioniziruyushchikh Izluchenii v Sel'skom Khozyaistve*. Kishinev, pp. 39–40.

254. Shvarts, K.K., Z.A. Grant, T.K. Mezhs and M.M. Grube. 1968. *Termolyuminestsentnaya Dozimetriya* (Thermoluminescent Dosimetry). "Zinatne" Publishers, Riga.

255. Engel', O.S. 1952. Vliyanie rentgenovskikh luchei na semena pshenitsy v zavisimosti ot stepeni ikh zrelosti (The effect of x-rays on wheat seeds depending upon their degree of maturity). *Dokl. AN SSSR*, vol. 85, No. 2, pp. 437–442.

256. Yudin, M.F. 1970. Dozimetriya Fotonnogo Izlucheniya (Dosimetry of Photon Radiation). "Standarty" Publishers, Moscow.

257. Yakobson, I.I. 1936. Deistvie gamma-luchei na khlopchatnik (The effect of gamma-rays on cotton). *Sots. Nauka i Tekhnika*, No. 4, pp. 61–64.

258. Yakobov, A.M. 1958. Deistvie gamma-luchei na energiyu prorastaniya i vskhozhesti semyan razlichnykh vidov khlopchatnika (The effect of gamma-rays on the germination energy and capacity of the seeds of different cotton varieties). *Dokl. AN UzSSR*, No. 7, pp. 51–55.

259. Yanchenko, K.V. 1959. K voprosu posledeistviya luchei Rentgena na semena pshenitsy (The problem of the after-effects of x-rays on wheat seeds). *Uchn. Zap. Krasnoyarsk. Gos. Ped. In-ta*, vol. 15, pp. 124–133.

260. Adams, I.D. and R.A. Nilan. 1958. After-effects of ionizing radiation in barley. II. Modification of x-irradiation of seeds in different concentrations of oxygen. *Radiation Res.*, vol. 8, pp. 111–115.

261. Alexander, L.T. 1959. Radioactive materials as plant stimulants—Field results. *Agron. J.*, vol. 42, pp. 252–255.

262. Altman, V., D. Rochlin and E. Gleichgewicht. 1923. Über entwicklungsbeschleunigenden und entwickenngchemmenden Einfluss der Röntgenstrahlen. "Froscher. Geb. Röntgenstrahlen". Bd. 31, pp. 51–62.

263. *ASTM Bull.* 1959. No. 239 (ASTMD 1671–63).

264. Barnetsky, F. Eine Jahresperiodische Anderung der Strahlenempfindichkeit bei Geräte. *Naturwissenschaften*, 1967, Bd. 47, pp. 116–117.

265. Barton, L.V. 1961. Anatomy of seeds. In: *Effects of Ionizing Radiations on Seeds.* Vienna, IAEA, pp. 25–31.

266. Barton, L.V. 1961. Experimental seed physiology of Boyce Thompson Institute for plant research. *Compt. rend. Assoc. internat. essais semences.* No. 4, pp. 561–596.

267. Barton, L.V. and R. Helen. 1946. Effect of age and storage condition of seeds on the yield of certain plants. *Contribs. Boyce Thompson. Inst.*, vol. 14, pp. 243–247.

268. Barton, G. et al. 1958. Chemical Processing Wastes Recovering Fission Products. *Ing. Eng. Chem.* No. 50, pp. 212–213.

269. Bequerel, H. 1901. Some chemical effects produced by radium ray. *Compt. rend. Acad. Sci.*, vol. 133, pp. 709–713.

270. Bora, K.C. 1958. Comparative effects of varying doses of x-rays and fast neutrons on the growth, development and induction of cytological changes in tetraploid and heploid wheat. In: *Proc. 2nd UN Int. Conf. PUAE*, vol. 27, p. 314.

271. Chapman, I.M. 1963. Computer calculation of dose rates in two-legged ducts using the albedo concept. *Trans. Amer. Nucl. Soc.*, vol. 6, No. 2, pp. 437–438.

272. Chapman, I.M. and C.M. Huddleston. 1966. Dose attenuation in two-

legged concrete ducts for various gamma-ray energies. *Nucl. Sci. Engng*. No. 25, pp. 66–68.

273. Coldera, P.G. 1972. Gamma-stimulation of potato tubers yor. *Stimulation Newsletter*, No. 2, pp. 5–11.
274. Czepa, A. 1924. The problem of growth promotion and functional increase caused by x- and radium-rays. *Strahlentherapie*, Bd. 16, pp. 913–918.
275. Davies, D.R. and E.T. Wall. 1961. Gamma-radiation and interspecific incompatibility in plants. In: *Effects of Ionizing Radiation on Seeds*. Vienna. IAEA, pp. 83–90.
276. Demmerques, P. Action des rayons gamma sur les bourgeons de la variete de poiriermax Red Bartlett. *Ibid*., pp. 581–592.
277. Desai, B.M. and B.K. Gaur. 1972. Effects of Low Doses of x-rays on Radish and Carrot. *Stimulation Newsletter*, No. 2, pp. 27–31.
278. Drachowska, M. and K. Sandero. 1961. Sb. Vysoke skyly Chek-technul Prare. *Paravian technul*., vol. 5, No. 1, pp. 25–30.
279. Evler, G. 1906. Ueber die heilende Wirkung der Röntgenstrahlen bei abgegrenzten Eiterungen. *Jahrb. J. Wissensch. Botanik*. Bd. 5, No. 56, pp. 416–421.
280. Fischnich, O., C. Pätzold and F. Heilinger. 1961. Influence of low doses of irradiation (X-rays and gamma rays of cobalt-60 on potato seeds). In: *Effects of Ionizing Radiation on Seeds*. Vienna, IAEA, pp. 553–565.
281. Geiger, C.S. 1906. The effect of the rays of radium on plants. *Proc. Amer. Assoc. Advance Sci*., vol. 55, pp. 326–331.
282. Geiger, C.S. 1907. Radioactivity a factor in plant environment. *Science*, vol. 25, pp. 263–264.
283. Geiger, C.S. 1907. Some effects of radioactivity on plants. *Ibid*., p. 264.
284. Geiger, C.S. 1908. Effects of radium rays on mitosis. *Ibid*., vol. 27, pp. 336–338.
285. Geiger, C.S. 1908. Effects of the rays of radium in plants. *Mem. N.Y. Botan. Garden*, vol. 4, pp. 1–78.
286. Geiger, C.S. 1909. The influence of radium rays on a few type processes of plants. *Popul. Sci. Monthly*, vol. 74, pp. 222–223.
287. Geiger, C.S. 1916. Present status of the problem of the effect of radium rays on plant life. *Mem. N.Y. Botan. Garden*, vol. 6, pp. 153–166.
288. Glubrecht, H. 1968. The present status and future aspects of radio-stimulation in crop plants in Europe. Vienna, FAO/IAEA (Rotoprint).
289. Glubrecht, H. 1966. *Kerntechnik*, Bd. 8, No. 11, p. 423.
290. Guilleminot, H. 1907. Effects of radium on the seeds and development of plants. *Arch. Leck. Med*., vol. 15, pp. 592.
291. Guilleminot, H. 1907. Effect of radium radiation and x-rays on germination. *Compt. rend. Assoc. franc. avavc Sci*., vol. 37, pp. 1344–1353.

292. Guilleminot, H. 1910. Effect of new radiations on plant. *J. belge. radiol.*, vol. 4 (4), pp. 537–547.

293. Guilleminot, H. 1901. Persistence de l'action des rayons x et du radium sur la grain a l'etat de lane latente. *Compt. Rend. Soc. biol.*, vol. 68, pp. 309–311.

294. Gustafsson, A. 1946. The x-ray resistance of dormant seeds of some agricultural plants. *Hereditas*, vol. 30, pp. 166–178.

295. Gustafsson, A. 1946. Effect of x-rays upon agricultural seeds. *Nature*, vol. 158, pp. 488–491.

296. Hagberg, A. and N. Nybom. 1954. Reaction of potatoes to x-irradiation and radio-phosphorus. *Acta Agric. Scand.*, vol. 4, pp. 578–584.

297. Iven, H. 1925. Neu Untersuchungen über die Wirkung Röntgenstrahlen auf Pflanzen. *Strahlentheraphie*. Bd. 19, No. 3, pp. 413–418.

298. Jamado, M. 1917. On the effect of Roentgen rays upon seeds of *oryza sativa. J. Phys. Therapy*, No. 6, refer. *Journ. Col. Agric. Univ. Tokyo*, vol. 8, No. 2, pp. 32.

299. Jonson, E.L. 1938. Growth of wheat plants from irradiated dry and soaked grains. *J. Colo Wyo. Acad. Sci.*, vol. 2(4), pp. 24–25.

300. Jonson, E.L. 1939. Floral development of certain species as influenced by the x-radiation of buds. *Plant Physiol.*, vol. 14, pp. 783–795.

301. Jonson, E.L. 1936. Susceptibility of seventy species of flowering plants to x-radiation. *Plant Physiol.*, vol. 11, pp. 319–342.

302. Köernike, M. 1904. wirkung von Röntgenstrahlen auf Keimung und Wachstum. *Ber. d. deutsch. bot. Ges.*, Bd. 22, pp. 148–155.

303. Köernike, M. 1905. Weitere Untersuchungen über die Wirkung von Röntgenund Radiumstrahlen auf die Pflanzen. *Ibid.*, Bd. 23, pp. 324–333.

304. Köernike, M. 1915. Über die Wirkung ferschiedenen starker Röntgenstrahlen auf Keimung und Wachstum bei den höheren Pflanzen. *Jahrb. f. Wiss. Botan.* Bd. 55, p. 416.

305. Köernike, M. 1916. Die biologische Einwirkungen der Röntgenstrahlen auf die Pflanzen. *Zeitschrift. Physik und diätet. Therapie*, No. 6, pp. 363–364.

306. Köernike, M. 1919. Die Wirkung der Röntgenstrahlen auf die Pflanzen. *Forscher Geb Röntgenstr.* Bd. 27, No. 10, p. 661.

307. Köernike, M. 1920. Wirkung der Röntgenstrahlen und pflanzen. *Ibid.*, vol. 27, pp. 661–670.

308. Kersten, H.J. and G.T. Smith. 1942. Failure of root tips of tomato seedlings germinated from x-rayed seeds to growth *in vitro. Plant Physiol.*, vol. 17, pp. 321–323.

309. Klingmüller, W. 1961. Radiation damage in *Vicia faba* seeds. In: *Effects of Ionizing Radiation on Seeds*. Vienna, IAEA, pp. 67–75.

310. Komuro, H. 1919. The effects of x-rays on the germination of *Oryza sativa. Bot. Mag.*, vol. 33, pp. 223–227.

311. Komuro, H. 1922. On the effect of Röntgen rays upon the growth of *Oryza sativa*. *Ibid.*, vol. 36, pp. 15–17.

312. Komuro, H. 1925. Physiological and cytological changes caused by hard and soft X-rays in *Vicia faba, Pisum sativum*. *Ibid.*, vol. 39, pp. 233–236.

313. Komuro, H. 1924. Studies on the effect of Röntgen rays upon the germination of *Oryza sativa*. *Ibid.*, vol. 38, pp. 1–22.

314. Komuro, H. 1923. Studies on the effect of Röntgen rays upon the development of *Vicia faba*. *J. Coll. Imp. Univ. Tokyo*, vol. 8, pp. 253–292.

315. Konzak, C.T. and W.R. Singleton. 1952. The relationship of polyploidy to the effect of thermal neutron exposure on plant. *Genetics*, vol. 57, pp. 596–597.

316. Le Doux, I.C. and A.B. Chilton. 1961. Gamma-ray streaming through two-legged rectangular ducts. *Nucl. Sci. Engg.*, vol. 11, No. 4 pp. 362–363.

317. Long, T.P. and H. Karsten. 1936. Stimulation of the growth of soybeans by soil x-rays. *Plant Physiol.*, vol. 11, No. 3, pp. 615–621.

318. Long, T.P. and H. Karsen. 1937. Structural changes produced in beatt tissues of soybean plants by the irradiation of dry seeds with soft x-rays. *Ibid.*, vol. 12, No. 1, pp. 191–195.

319. MacKey, I. 1951. Neutron and x-ray experiments in barley. *Hereditas*. vol. 37, pp. 421–464.

320. MacKey, I. 1952. The biological action of x-rays and fast neutrons on barley and wheat. *Arch. Botanick*, vol. 1, pp. 545–556.

321. Maldinay, M., Touvenin. 1898. De l'influence des rayons X-sur la germinations. *Rev. Gen.*, vol. 10, pp. 81–86.

322. Meletti, P., D'Amato F. 1961. The embryo transplantation technique in the study of embryoendosperm relations in irradiated seeds. In: *Effects of Ionizing Radiations on Seeds*. Vienna, IAEA, pp. 47–57.

323. Menyhert, Z. 1970. Radiation stimulation of tomatoes. *J. Stimulation Newsletter*, No. 1, pp. 18–20.

324. Menyhert, Z. 1967. The utilization in agriculture of the stimulating effect of low dose ionizing radiations. *Atomtechnikal Tagekortato*, vol. 10, pp. 187–201.

325. Mericle, Z.W. and R.P. Mericle. 1957. Irradiation of developing plant embryos. I. Effects of external irradiation (x-rays) on barley embryogeny, germination and subsequent seedling development. *Amer. J. Bot.*, vol. 44, pp. 9–13.

326. Micke, A. 1961. Comparison of the effects of x-rays and thermal neutrons on viability and growth of sweet clover (melilotus albus) after the irradiation of dry seeds. In: *Effects of Ionizing Radiation on Seeds*. Vienna, IAEA, pp. 403–409.

327. Miege, E. and H. Coupe. 1914. De l'influence des rayons sur la vegetation. *C.R. Acad. Sci. Paris*, vol. 159, No. 4, pp. 338–340.
328. Nilan, R.A. et al. 1961. The oxygen effects in barley seeds. In: *Effects of Ionizing Radiation on Seeds*. Vienna, IAEA, pp. 139–145.
329. Notam, N.K. A study of differences in the radiosensitivity of some inbreds and hybrids in maize. *Ibid*., pp. 475–485.
330. *Nucleonics*, 1965, vol. 8, No. 23.
331. *Nucleonics*, 1966, vol. 5, No. 24.
332. Nybom, N. et al. 1953. Biological effects of x-irradiation at low temperatures. *Heredity*, vol. 30, pp. 445–457.
333. Palenzona, D.Z. 1961. Effects of high doses of x-ray on seedling growth in wheat of different ploidy. In: *Effects of Ionizing Radiation on Seeds*. Vienna, IAEA, pp. 533–540.
334. Pannonhalmi, K. 1970. Stimulation of emergence by x-irradiation in sugar beets of different grades of ploidy, *J. Stimulation Newsletter*, No. 1, pp. 22–23.
335. Patten, R.E.P. and S.B. Wigoder. 1929. Effect of x-ray on seeds. *Nature*, vol. 123, pp. 606–611.
336. Simon, I., Z. Menyhert, K. Pannonhalmi, A. Balint and J. Pal. 1968. Preliminary radio-stimulation experiments in Hungary and the perspectives of stimulation. Vienna. FAO/IAEA (Rotaprint).
337. Prevotel, M. Elan-2B, pilote industrial pour la fabrication de sources au cesium-137 et au strontium-90. *Industries atomiques*, 1970, vol. 14, No. 9/10, p. 41.
338. Promcy, G. and M. Drevon. 1912. Effect of x-rays on germination. *Rev. Gen. bot.*, vol. 24, No. 281, pp. 177–195.
339. Randolf, M.L. and A.H. Haber. 1961. Production and decay of free radicals induced by x-irradiation of dry lettuce seed. In: *Effects of Ionizing Radiations on Seeds*, Vienna, IAEA, pp. 57–63.
340. Rizzo, F.X., C. Eisenhaner and A. Quadrado. 1960. Attenuation of cobalt-60 gamma-radiation through air ducts in concrete shields. *Trans. Amer. Soc.*, vol. 3, No. 2, pp. 562–563.
341. Rochlin, E. 1925. Zur Frage vom biologischen Einfluss der Röntgenstrahlen. *Fortschr. Geb. Röntgenstrahlen*, Bd. 33, p. 971.
342. Saric, M. 1958. The dependence of irradiation effects in seeds on the biological properties of the seed. In: *Proc. Ind UN Int. conf. PUAE*, vol. 27, pp. 223–229.
343. Saric, M. 1961. The effects of irradiation in relation to the biological traits of the seed irradiated. In: *Effects of Ionizing Radiation on Seeds*. Vienna, IAEA, pp. 183–187.
344. Sax, K. 1950. The Cytological Effects of Low-intensity Radiation. *Science*, vol. 112, pp. 332–337.
345. Schmidt, H. 1910. Experimentelle Untersuchungen über die Wirkung

kleinerer und grösserer Röntgenstrahlenmengen auf junge Zellen. *Berl. Klin. Wochenschr.*, Bd. 47, No. 21, pp. 972–974.

346. Schober, A. 1896. An experiment with x-rays and germinating plants. *Berl. deut. bot. ges.*, Bd. 14, pp. 108–110.

347. Schwanitz, H. 1972. Investigation on delayed effects of ionizing radiation stimulation effects on young kale plants (*Brassica oleracer*, L. *Acephala* D.C.) after exposure of seeds of the parental generation to a high dose of x-rays. *J. Stimulation Newsletter*, No. 2, pp. 15–20.

348. Schwarz, E. 1913. Der Wachstumreise der Röntgenstrahlen suf pflanzliche Gewebe. *Münchener med. Wochenschr.*, No. 39, pp. 2165–2169.

349. Schwartz, G.A., A. Czepa and H. Schindler. 1925. Recent studies on the effect of x-rays on plant. *Strahlentherapie*, Bd. 20, pp. 210–216.

350. Shull, C. and G.W. Mitchell. 1933. Stimulative effects of x-rays on plant growth, *Plant Physiol.*, vol. 8, pp. 287–296.

351. Simon, A. and C. Clifford. 1956. The attenuation of neutrons by air ducts in shields. *Nucl. Sci. Engg.*, No. 1, pp. 156–157.

352. Sparrow, A.H. et al. 1961. Some factors affecting the responses of plants to acute and chronic radiation exposures. In: *Effects of Ionizing Radiations on Seeds.* Vienna, IAEA, pp. 133–187.

353. Sparrow, A.H. and N.R. Singleton. 1953. The use of radiocobalt use source of gamma rays and some effects of chronic irradiation on growing plants. *Amer. Naturalist*, vol. 87, pp. 29–48.

354. Stan, S. and A. Croitoru. 1970. Effects of low, moderate and high levels of gamma radiation (^{60}Co) on soybean plants in an analysis of growth and yield. *J. Stimulation Newsletter*, No. 1, pp. 23–27.

355. Stein, G. and R. Richter. 1961. The effects of x-ray irradiation in conjunction with red and far-red light on lettuce seed germination. In: *Effects of Ionizing Radiation on Seeds.* Vienna, IAEA, pp. 197–204.

356. Süb, A. and W. Grobe. 1967. Aufbau und Betrieb der Bestrahlungsanlage Rotatron. *Kerntechnic*, Bd. 9, p. 302.

357. Süs, A. and B. Bretschneider-Hermann. 1972. The effects of low radiation doses on barley and wheat. *J. Stimulation Newsletter*, No. 2, pp. 36–45.

358. Süs, A. 1965. Die Wirkung Kleiner Strahlendosen bei Saatgutbestrahlung. *Bayerisches landwirtschaftliches Jahrbuch*, No. 1, pp. 42–55.

359. Terrell, C.W. and A.I. Jerri. 1962. Measured Radiation Distribution in Shelter Entranceways. *Trans. Amer. Soc.*, vol. 5, No. 2, pp. 393–394.

360. Warner, B.F. 1957. La fabrication de sources de ^{137}Cs d'un kilocurie. *Energia nucl.*, vol. 1, No. 3, pp. 161–167.

361. Uber, F. and F. Goodspeed. 1934. Influence of death criteria of the x-ray survival curves of the fungus Neurospora. *J. Gen. Physiol.*, vol. 17, pp. 577–590.

362. Vidal, P. 1959. Angmentation du rendement et des qualites des

Cultures par irradiation des Semences. *Industries Atomiques*, vol. 1/2. p. 34.

363. Wetterer. 1911. Beitrag zur Kenntnis der biologischen Wirkung der Röntgenstrahlen auf das Wachstum der Pflanzen. *Münch. med. Wochenschr.*, No. 42, pp. 2237–2242.

364. Wolf, S. and A.M. Sicard. 1961. Post-irradiation storage and the growth of barley seedlings. In: *Effects of Ionizing Radiations on Seeds.* Vienna, IAEA, pp. 171–179.